Environmental Fate of Emerging Organic Micro-Contaminants

Environmental Fate of Emerging Organic Micro-Contaminants

Special Issue Editors

Peter S. Hooda
John Wilkinson

MDPI • Basel • Beijing • Wuhan • Barcelona • Belgrade

Special Issue Editors

Peter S. Hooda   John Wilkinson

Kingston University London   University of York

UK   UK

Editorial Office

MDPI

St. Alban-Anlage 66

4052 Basel, Switzerland

This is a reprint of articles from the Special Issue published online in the open access journal *Applied Sciences* (ISSN 2076-3417) in 2019 (available at: https://www.mdpi.com/journal/applsci/special_issues/Micro_Contaminants).

For citation purposes, cite each article independently as indicated on the article page online and as indicated below:

LastName, A.A.; LastName, B.B.; LastName, C.C. Article Title. *Journal Name* **Year**, *Article Number*, Page Range.

ISBN 978-3-03921-367-2 (Pbk)

ISBN 978-3-03921-368-9 (PDF)

Contents

About the Special Issue Editors

Peter Hooda, Ph.D. (London), is currently Associate Professor (land and water environmental science) at Kingston University London. He is a fellow of the Royal Society of Chemistry (FRSC) and the British Soil Science Society (FBSSS). Prior to joining Kingston University, Peter worked at several leading environmental science research laboratories, including the James Hutton Institute (Aberdeen, Scotland) and Lancaster University. Much of his work has focused on land and water contamination issues, covering both rural and urban sources of soil/water pollution. This includes a range of contaminants, including nutrients, potentially toxic trace elements, and organic micro-contaminants. Peter is a seasoned editor, and his portfolio includes Trace Elements in Soils (Wiley-Blackwell, 2010), Microbial and Nutrient Contaminants in Fresh and Coastal Waters (*Journal of Environmental Management* 87, 2008) and Heavy Metals in Soils (*Adv. Environ. Res.* 2003). Peter currently serves as an Associate Editor with the *Journal of Environmental Management*, and sits on the Editorial Board of several journals, including *Science of the Total Environment*, *Frontiers in Environmental Science*, and *Applied Sciences* (MDPI).

John Wilkinson is an environmental toxicologist and analytical chemist in the Department of Environment and Geography at the University of York (United Kingdom). He has worked on several large international projects, including the European iPiE project on the intelligent assessment of pharmaceuticals in the environment, the British Council STREAM project on the safe use of treated wastewater for crop irrigation in Israel and Palestine, the MRC drivers of human exposure to antibacterial resistance in the Sri Lankan environment project, among others. John currently co-coordinates the Global Monitoring of Pharmaceuticals project, which comprises over 80 partner institutions aiming to assess pharmaceutical contamination of rivers, on a global scale, for the first time. Before joining the University of York, John completed his PhD at Kingston University London, investigating the occurrence, bioaccumulation, fate, and distribution of pharmaceuticals, plasticisers, illicit drugs, and perfluorinated compounds in the aquatic environment. John has also spent time investigating emerging eco-human health issues of endocrine disrupting chemicals at the University of Colorado Denver (Colorado, USA).

Editorial

Special Issue on the Environmental Fate of Emerging Organic Micro-Contaminants

John Wilkinson [1] and Peter S. Hooda [2,*]

[1] Environment and Geography Department, University of York, York YO10 5NG, UK
[2] Faculty of Science, Engineering and Computing-Kingston University, Kingston-upon-Thames KT1 2EE, UK
* Correspondence: p.hooda@kingston.ac.uk

Received: 19 July 2019; Accepted: 24 July 2019; Published: 26 July 2019

1. Introduction

The toxicity and fate of pharmaceuticals and other emerging micro-organic contaminants in the natural and built environments have been the focus of much research over the last 20 years [1,2]. Recently, particular focus has been centred on the fate of antimicrobial chemicals, including antibiotics and antifungals, as well as other medicinal and anthropogenic chemicals [3]. The occurrence of such contaminants in the environment is thought to contribute to adverse outcomes, including the selection of antimicrobial resistance [4] and endocrine disruption in exposed non-target organisms [5]. Much work is needed to elucidate the environmental fate of micro-organic contaminants, particularly related to their removal from sewage effluents and their potential uptake into both aquatic and terrestrial foodstuffs [2]. Underpinning such work is the need for innovative and robust analytical methodologies for the quantification of these contaminants in a broad range of environmental matrices.

2. Micro-Organic Contaminants: Perspectives on Detection, Effects and Removal/Treatment

This Special Issue (SI) brings together a broad range of recent advances in the field of micro-organics, ranging from medicinal contaminants to industrial chemicals in the environment. Notably, these range from chemical extraction and large-scale analysis to adverse effects on non-target aquatic organisms and potential risk to humans via contaminated foodstuffs. Additionally, this Special Issue presents novel contaminant treatment and degradation methods of both physical and biological natures. A total of nine articles were selected for publication.

The first article, authored by Svahn and Björklund, presents an innovative and widely accessible methodology by which pharmaceuticals can be extracted from complex environmental matrices using a household espresso machine [6]. Such advancements have the potential to increase accessibility to environmental research, which is dominated by the use of complex and often prohibitively expensive equipment. As such, access to technology, such as the highly sensitive mass spectrometers needed to quantify micro-organic contaminants, tends to disproportionally favour wealthy countries (e.g., Western Europe, North America and China) [7] and institutions. The second article of this SI, authored by Wilkinson et al., presents an innovative methodology by which access to such analytical equipment may be achieved [7]. The authors accomplish this via affordable, validated and scientifically robust means for quantification of a broad suite of antimicrobials and other common medicinal products in a worldwide water enabling study [7].

With the capability and means to extract and quantify micro-organic contaminants, the study of the concomitant biological effects they may elicit in exposed organisms was also a focus of this SI. The selection of resistance to antimicrobial medicines is of increasing concern to the scientific community. In the third article of this SI, Osińska et al. demonstrate that the occurrence of antimicrobial-resistant genes to beta-lactams and tetracyclines is frequently detected and poorly, if at all, removed in wastewater treatment plants [8]. Similarly, the authors of the fourth article in this SI, Park et al., found that contaminants commonly detected in wastewater (plasticisers and a biocide) manipulate the expression of the egg yolk protein vitellogenin in a compound and dose-specific manner [9]. Studies elucidating the potential biological impacts of micro-organic contaminants were not limited to lower organisms in this SI. The fifth article presents a human health risk assessment for secondary exposure to veterinary medicines via consumption of contaminated shrimp [10]. Tsai et al. conclude that while some aqua-farmed shrimp in Taiwan do bioaccumulate detectable concentrations of antimicrobials, the risk to humans consuming them is negligible.

Ultimately, to reduce the risk of biological disruption on any level, removal or degradation of micro-organic contaminants is key. In total, four articles in this SI relate to various approaches to degrading or removing micro-organic contaminants. The sixth article of this SI, authored by Lee et al., establishes that the types of perfluoro-alkyl contaminants entering wastewater treatment are linked to the type of local industry, and their removal is most effective via a combination of oxidation and activated carbon adsorption [11]. Similarly, the seventh article, authored by Baresel et al., reinforces the finding that adsorption of some micro-organic contaminants is an effective means to separate them from wastewater effluent prior to discharge into the environment [12]. In addition to physical separation/degradation, the last two articles of this SI focus on biological removal of such contaminants. Specifically, Zhang et al. demonstrate effective and rapid degradation of a common industrial contaminant using the Gram-positive bacterium *Rhodococcus* sp. [13]. This significant finding raises the potential for this bacterium to be used for targeted remediation in the environment or for treatment of industrial wastes [13]. Similarly, the last article of this SI, authored by Bai et al., shows effective decolorization of various dyes by the Gram-negative bacterium *Pseudomonas putida*, again indicating the potential of biological treatments in the breakdown of micro-organic contaminants [14].

3. Perspectives on the Future of Research on Micro-Organic Contaminants in the Environment

While the work presented in this SI offers concrete advancements to the study of anthropogenic contaminants in the environment, much work is still needed. In future years, it can be anticipated that increasing water scarcity in light of global climate change and increasing urbanisation of the global population will stress existing resources for clean freshwater. With this stress may come increasing contamination of water. Furthermore, little is currently known about the occurrence of micro-organic pollutants in much of the developing world. Effective means of risk-based prioritisation of contaminants, particularly in foodstuffs and bathing and drinking water, are needed within this context. Ultimately, much additional work is needed to identify affordable and effective means by which to remove these contaminants from water.

References

1. Wilkinson, J.L.; Hooda, P.S.; Barker, J.; Barton, S.; Swinden, J. Ecotoxic pharmaceuticals, personal care products, and other emerging contaminants: A review of environmental, receptor-mediated, developmental, and epigenetic toxicity with discussion of proposed toxicity to humans. *Crit. Rev. Environ. Sci. Technol.* **2016**, *46*, 336–381. [CrossRef]
2. Wilkinson, J.; Hooda, P.S.; Barker, J.; Barton, S.; Swinden, J. Occurrence, fate and transformation of emerging contaminants in water: An overarching review of the field. *Environ. Pollut.* **2017**, *231*, 954–970. [CrossRef]
3. Shao, S.; Hu, Y.; Cheng, J.; Chen, Y. Research progress on distribution, migration, transformation of antibiotics and antibiotic resistance genes (ARGs) in aquatic environment. *Crit. Rev. Biotechnol.* **2018**, *38*, 1195–1208. [CrossRef] [PubMed]
4. Tell, J.; Caldwell, D.J.; Häner, A.; Hellstern, J.; Hoeger, B.; Journel, R.; Mastrocco, F.; Ryan, J.J.; Snape, J.; Straub, J.O.; et al. Science-based targets for antibiotics in receiving waters from pharmaceutical manufacturing operations. *Integr. Environ. Assess. Manag.* **2019**, *15*, 312–319. [CrossRef] [PubMed]
5. Windsor, F.M.; Ormerod, S.J.; Tyler, C.R. Endocrine disruption in aquatic systems: Up-scaling research to address ecological consequences. *Biol. Rev.* **2018**, *93*, 626–641. [CrossRef] [PubMed]

6. Svahn, O.; Björklund, E. Extraction Efficiency of a Commercial Espresso Machine Compared to a Stainless-Steel Column Pressurized Hot Water Extraction (PHWE) System for the Determination of 23 Pharmaceuticals, Antibiotics and Hormones in Sewage Sludge. *Appl. Sci.* **2019**, *9*, 1509. [CrossRef]

7. Wilkinson, J.L.; Boxall, A.; Kolpin, D.W. A novel method to characterise levels of pharmaceutical pollution in large-scale aquatic monitoring campaigns. *Appl. Sci.* **2019**, *9*, 1368. [CrossRef]

8. Osińska, A.; Korzeniewska, E.; Harnisz, M.; Niestępski, S. Quantitative Occurrence of Antibiotic Resistance Genes among Bacterial Populations from Wastewater Treatment Plants Using Activated Sludge. *Appl. Sci.* **2019**, *9*, 387.

9. Park, K.; Jo, H.; Kim, D.K.; Kwak, I.S. Environmental Pollutants Impair Transcriptional Regulation of the Vitellogenin Gene in the Burrowing Mud Crab (Macrophthalmus Japonicus). *Appl. Sci.* **2019**, *9*, 1401. [CrossRef]

10. Tsai, M.Y.; Lin, C.F.; Yang, W.C.; Lin, C.T.; Hung, K.H.; Chang, G.R. Health Risk Assessment of Banned Veterinary Drugs and Quinolone Residues in Shrimp through Liquid Chromatography–Tandem Mass Spectrometry. *Appl. Sci.* **2019**, *9*, 2463. [CrossRef]

11. Lee, S.H.; Cho, Y.J.; Lee, M.; Lee, B.D. Detection and Treatment Methods for Perfluorinated Compounds in Wastewater Treatment Plants. *Appl. Sci.* **2019**, *9*, 2500. [CrossRef]

12. Baresel, C.; Harding, M.; Fång, J. Ultrafiltration/Granulated Active Carbon-Biofilter: Efficient Removal of a Broad Range of Micropollutants. *Appl. Sci.* **2019**, *9*, 710. [CrossRef]

13. Zhang, Y.; Ji, J.; Xu, S.; Wang, H.; Shen, B.; He, J.; Qiu, J.; Chen, Q. Biodegradation of Picolinic Acid by Rhodococcus sp. PA18. *Appl. Sci.* **2019**, *9*, 1006. [CrossRef]

14. Bai, Z.; Sun, X.; Yu, X.; Li, L. Chitosan Microbeads as Supporter for Pseudomonas putida with Surface Displayed Laccases for Decolorization of Synthetic Dyes. *Appl. Sci.* **2019**, *9*, 138. [CrossRef]

Article

Extraction Efficiency of a Commercial Espresso Machine Compared to a Stainless-Steel Column Pressurized Hot Water Extraction (PHWE) System for the Determination of 23 Pharmaceuticals, Antibiotics and Hormones in Sewage Sludge

Ola Svahn * and Erland Björklund

Department of Environmental Science and Bioscience, Faculty of Natural Science, Kristianstad University,
SE-291 39 Kristianstad, Sweden; erland.bjorklund@hkr.se
* Correspondence: ola.svahn@hkr.se

Received: 26 February 2019; Accepted: 4 April 2019; Published: 11 April 2019

Abstract: Two green chemistry extraction systems, an in-house stainless-steel column Pressurized Hot Water Extraction system (PHWE) and a commercially available Espresso machine were applied for analysing 23 active pharmaceutical ingredients (APIs) in sewage sludge. Final analysis was performed on UPLC-MS/MS using two different chromatographic methods: acid and basic. When analysing all 23 APIs in sewage sludge both extraction methods showed good repeatability. The PHWE method allowed for a more complete extraction of APIs that were more tightly bound to the matrix, as exemplified by much higher concentrations of e.g., ketoconazole, citalopram and ciprofloxacin. In total, 19 out of 23 investigated APIs were quantified in sewage sludge, and with a few exceptions the PHWE method was more exhaustive. Mean absolute recoveries of 7 spiked labelled APIs were lower for the PHWE method than the Espresso method. Under acid chromatographic conditions mean recoveries were 16% and 24%, respectively, but increased to 24% and 37% under basic conditions. The difference between the PHWE method and the Espresso method might be interpreted as the Espresso method giving higher extraction efficiency; however, TIC scans of extracts revealed a much higher matrix co-extraction for the PHWE method. Attempts were made to correlate occurrence of compounds in sewage sludge with chemical properties of the 23 APIs and there are strong indications that both the number of aromatic rings and the presence of a positive charge is important for the sorption processes to sewage sludge.

Keywords: espresso coffee machine extraction; pressurized hot water extraction; pharmaceuticals; antibiotics; hormones; sewage sludge; ion suppression; UPLC MS/MS; basic buffer

---

1. Introduction

More than 1000 different active pharmaceuticals ingredients (APIs) are today used in Sweden [1]. The release of APIs into the water environment has been a subject of research for more than 30 years [2], and their ubiquitous occurrence at varying concentration levels have been shown in wastewater, surface water, sediment, groundwater and drinking water [3–9]. Large research resources have been spent worldwide on investigating the occurrence of pharmaceutical residues in the water phase, a significantly smaller proportion resources of these compounds' presence in sewage sludge [10]. The production of sewage treatment plant (STP) sludge in Sweden is estimated at approximately 207,500 tons of dry matter, finalized at over 400 STPs. The sludge contains, besides carbon, about 3% phosphorus and 3.5% nitrogen. This means that around 6000 tons of phosphorus and 7000 tons of nitrogen can be recycled in Sweden and returned to the soil via sludge each year [11]. Sludge spread on farmland is the largest single use category, since it is the most economical outlet for sludge and offers the opportunity to recycle plant nutrients and organic matter to soil for crop production. Agricultural use is estimated at approximately 50,000 tons, which corresponds to 25% of the total Swedish net production [12]. Sludge spreading in Sweden is regarded as an environmentally hazardous activity, but is still not subject to authorization or reporting, though authorities may impose stricter requirements in individual cases. From a European perspective the European Council Directive 86/278/EEC on the protection of soil, regulates sewage sludge use in agriculture in the EU. In several EU countries Directive 86/278/EEC is complemented by national legislation on soil protection; however, such legislation does not include regulation of pharmaceutical residues, even though there is increasing societal awareness and debate on the fate of APIs during sludge management. The fact that the chemical load in the form of pharmaceuticals and most other emerging contaminants (ECs), is unregulated both in wastewater effluents and in produced sewage sludge at the STPs, may in part be due to the fact that sludge chemically is a very complex matrix, which severely complicates the chemical analysis [13,14]. Nevertheless, sewage sludge is a sink for many of these substances [15] and given the spreading of sludge on farmland makes it important to seek valid results on its content of APIs. Traditionally, a variety of techniques have been used for extracting organic pollutants, including APIs, from sewage sludge such as Soxhlet, ultrasound assisted extraction, microwave-assisted extraction, mechanical shaking, supercritical fluid extraction and pressurized liquid extraction [16–20]. Since APIs hold a variety of physicochemical properties it is difficult to find a single method capable of analysing such a large number of differing chemical substances. Furthermore, emerging in recent years, there is a growing environmental concern in chemistry reshaping this field towards "green chemistry" [21], which is defined as the design of chemical products and processes that reduce or eliminate the use or generation of hazardous substances [22]. One possible technique possessing properties with the potential of meeting both these criteria is pressurized hot water extraction (PHWE) [23].

From a scientific perspective, experimental work on PHWE dates back to at least 1994 when Hawthorne et al. [24], interested in finding environmentally friendly extraction methods, suggested water as a clean solvent for the extraction of non-polar analytes from environmental samples. In their pioneering work, performed on their in-house constructed equipment using stainless-steel columns, they found an increase in solubility of non-polar organics such as polyaromatic hydrocarbons (PAHs) in water with increasing temperature. The basic experimental set-up presented by Hawthorne and co-workers [24] has previously been used by us for the extraction of APIs from sediment [9], using a stainless-steel column Pressurized Hot Water Extraction system (PHWE method). This system and methodology was applied in this work as a reference methodology using a temperature of 150 °C as it has previously been shown that temperatures exceeding 100 °C has a very positive effect on extraction efficiency for pharmaceuticals in solid matrices [25]. Yet, too high temperatures (>200 °C) may cause thermal degradation of some compounds [26] and increased ion suppression as a consequence of severe co-extraction of unwanted components. A temperature of 150 °C is in our experience a suitable balance between good extraction efficiency combined with relatively moderate co-extraction of undesired matrix components.

One of the world's most common commercial pressurized hot water extraction apparatuses has been in use since the early 20th century. After the Second World War and in the early 1960s this extraction technique, the espresso machine, underwent technical improvement with the sole purpose of delivering great coffee by the extraction of flavours from grinded coffee beans mounted in the porta filter. Typically, modern espresso machines reach temperatures around 100 °C and operates at pressures of 15–20 bars. Substantial research has been done to better understand what parameters govern the extraction process of coffee with the aim of achieving a balanced coffee aroma [27–30], but commercial espresso machines have also been used to determine different contaminants such as polyaromatic hydrocarbons in soil [31], airborne pesticides in particulate matter trapped in filters [32] and polychlorinated biphenyls in soil [33]. In this work, a commercially available espresso machine was used for the extraction of APIs holding a variety of physicochemical properties from dewatered stabilized sewage sludge. The efficiency and reproducibility of the espresso machine extraction (Espresso method) was evaluated and results were compared to results from our Pressurized Hot Water Extraction (PHWE) as well as literature data obtained with traditional extraction techniques and methods. Post-extraction analysis with UPLC-MS/MS were conducted in both acid and basic chromatographic environments, as this previously has been shown to have an effect on the level of matrix influence in the chromatographic run [34]. The key advantages of a successful espresso machine extraction (Espresso method) would be very simple packing and maintenance procedures and a very fast extraction process.

2. Materials and Methods

2.1. Chemicals and Reagents

Ultra-pure water (18.2 MΩ) was obtained from an OPTIMA water purification system (Elga Ltd., High Wycombe, Buckinghamshire, UK). Reference standards were purchased from Sigma-Aldrich Sweden AB (Stockholm, Sweden). A list of compound details is found in Table S1, while chemical formulas, structures and molar masses are easily available on the Internet at PubChem, Drugbank, and ChemSpider. Acetonitrile (ACN) and methanol (MeOH), both OPTIMA grade, used for the chromatographic mobile phase were purchased from Fisher Scientific (Gothenburg, Sweden). Formic acid (FoA), ammonium hydroxide solution (25% sol.), ammonium hydrogen carbonate, disodium ethylenediaminetetraacetate (Na_2EDTA), ascorbic acid, and ammonium hydroxide and silicon carbide were purchased from Sigma-Aldrich. Stabilised dewatered sewage sludge was obtained from Kristianstad STP (Scania, Sweden) and lyophilized in-house.

2.2. PHWE

The in-house constructed PHWE system is schematically shown in Figure 1.

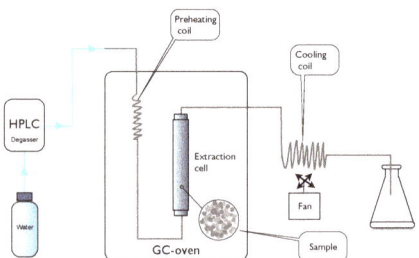

Figure 1. Schematic picture of the in-house stainless-steel column pressurized hot water extraction (PHWE) system.

The PHWE system consisted of a Waters Alliance 2690 HPLC system (Waters, Milford, MA, USA). In all experiments, the HPLC pump supplied water at pH 7, via stainless-steel tubing (1/16 in. o.d. and 0.040 in. i.d.), to the pressurized stainless-steel column (length 7.0 cm × i.d. 1.0 cm), in which the extraction took place. The extraction cell was placed inside a Varian 3400 GC oven (Walnut Creek, CA, USA) heated to 150 °C. A portion of 0.2 g lyophilized sludge was homogenized and mixed with 8.0 g of silicon carbide. A regular coffee filter (Melitta® 102, Melitta Europa GmbH & Co. KG, Minden, Germany) shaped into a cylinder covered the inside of the column and a circular shaped glass microfiber filter (GF/C 47 mm, Whatman, GE Healthcare UK Limited, Buckinghamshire, UK) was positioned at the outlet nut of the extraction column. The column was filled with the mixture of sludge and silicon carbide. The mixture was spiked at the top with 30 μL of an internal standard mixture containing 7 labelled standards (Table S1) and left to soak for 10 min. The content of the column was then packed with a stomp and the coffee filter edges folded and the column closed with an inlet nut.

The column was mounted inside the oven and connected to the stainless-steel tubing. The degasser was activated and the HPLC pump flow set to 1 mL/min. When the eluate emerged at the collection point the oven temperature was set to 150 °C and flow adjusted to 1.8 mL/min. A single run was completed in 13 min and gave 23 mL of extract. Pressure was monitored to between 11 and 12 bars. Additionally, the eluate was centrifuged at 5000 rpm for 10 min on an SIGMA 4-16KS Centrifuge (SIGMA Laboratoriezentrifugen, Osterode am Harz, Germany). Then a SPE protocol followed, as described below. The column was thereafter disconnected and the system was cleaned with a by-pass procedure at 90 °C for 5 min with the flow 1 mL/min. The column was emptied and reinstalled in the system and thereafter washed at 150 °C for 10 min at 2 mL/min. All experiments were conducted in triplicate.

2.3. Espresso Machine Extraction (Espresso Method)

A Rancilio Miss Silvia E Espresso Machine (Rancilia Group S.p.A., Villastanza di Parabiago, Italy) adopted for pods was used to conduct the espresso extractions. A portion of 0.2 g lyophilized sludge was homogenized and mixed with 8.0 g of silicon carbide. The mixture was spiked with 30 µL of internal standard mixture containing 7 labelled standards (Table S1) and left to soak for 10 min. A regular coffee filter cut in half enclosed the mixture and served as an extraction pod. At the bottom of the espresso porta filter a 30 mm 0.45 µm GF/C glass fibre filter was placed. The espresso machine was started when the ready lamp was lit and a total of 60 mL eluate was collected in a beaker. Previous Espresso extraction methods for environmental applications have used extraction volumes of 50 mL for sample sizes up to 5.0 g [31,32]. Our chosen volume of 60 mL was at the high-end of this scale and was also the maximum volume suitable for the SPE protocol, as described below. The cleaning procedure recommended by the supplier (placing 0.5 g of Espresso detergent in a blind filter and flushing for 10 s repeatedly) was performed 10 times in a row after each sample extraction was completed. The espresso experiments were conducted in triplicate.

2.4. Solid-phase Extraction (SPE Method)

A SPE robot RapidTrace+ (Biotage, Uppsala, Sweden) was used for *conditioning* and *eluting samples*, while a SPE manifold (Agilent, Santa Clara, CA, USA) was used for *loading/extracting* the samples. Samples were concentrated by an automatic evaporator TurboVap LV (Biotage, Uppsala, Sweden) using clean air produced by an Atlas Copco SF2 Oil-free air system (Atlas Copco Airpower n.v. B, Wilrijk, Belgium), delivering certified 100% oil-free air, complying with ISO 8573-1 CLASS 0 certification. Both the PHWE extracts and the Espresso extracts were concentrated and purified using 200 mg Oasis HLB SPE cartridges (Waters). The HLB SPE cartridges were conditioned with 5 mL of MeOH followed by 5 mL of reagent water prior to extraction. HLB SPE cartridges were mounted in the sample manifold together with a 70 mL plastic syringe container on top. Thereafter one volume of PHWE-or Espresso extract was added, together with 50 µL FoA (10%) and 50 µL saturated EDTA solution. Samples were then passed through the SPE cartridges at a rate of ca. 5 mL/min. Afterwards, the cartridges were air-dried under a positive pressure for 15 min, a step which proved crucial for both recovery results and evaporation time reproducibility, since drying the cartridges on the manifold was insufficient. SPE cartridges were mounted in the Rapid Trace+ SPE robot and analytes were eluted with 6 mL MeOH into disposable borosilicate glass tubes (PYREX®, 16 × 100 mm, Corning Incorporated, Corning, NY, USA). The extracts were evaporated to complete dryness (22 min) at 40 °C. The dry extracts were reconstituted to a total volume of 1 mL by adding 100 µL MeOH to the dry borosilicate glass tube, followed by a rapid swirling (vortexing) for 15 s. Thereafter 885 µL water was added followed by 15 s of rapid swirling, and transferred to a vial. Finally, 10 µL of *instrumental standard solution* (Thiacloprid-d4, 250 pg/µL, Table S1) was added along with 5 µL saturated EDTA solution. A volume of 1 µL of the final sample was injected into the UPLC-ESI-MS/MS system. The same SPE procedure was used for both extraction techniques.

2.5. Analytical Separation and Detection (UPLC-MS/MS Method)

The liquid chromatographic system, the mass spectrometer and the analytical methodology have all previously been described in detail [34,35]. In short, the UPLC-MS/MS system used was a Waters Acquity UPLC H-Class connected to a Xevo TQ-STM triple quadrupole mass spectrometer (Waters Micromass, Manchester, UK), equipped with a Z-spray electrospray interface. The UPLC H-Class consisted of a Quaternary Solvent Manager (QSM), a Sample Manager with Flow-Through Needle (SM-FTN) and a Column Manager (CM) enabling fast column switching between two different columns, running at two different pH (Waters, Milford, MA). Two Acquity UPLC BEH C18 columns (2.1 mm i.d. × 50mm, 1.8 mm) in parallel, maintained at 40 °C, were installed in the column manager. Nitrogen was used as both drying gas and nebulizer gas delivered by an Infinity Nitrogen Generator (Peak Scientific Instruments Ltd., Inchinnan, UK). For operation in MS/MS mode, the collision gas was argon 99.995% (AGA Gas AB, Malmö, Sweden). Waters UNIFI software (Version 1.7) controlled the UPLC-MS/MS system.

2.6. Calculations

2.6.1. Absolute Recoveries of Isotopic Labelled Compounds

We calculated the absolute recovery expressed as a percentage of the recovery of each isotopic labelled standard. The absolute recovery was then calculated according to Equation (1):

$$\text{Absolute Recovery} = 100 \cdot (A_n)/(A_l) \tag{1}$$

A_n = the area of the daughter m/z for the labelled compound in spiked sample from either the *PHWE* extracts or the *Espresso* extracts. A_l = the area of the daughter m/z for the labelled compound in standard solution.

2.6.2. Quantification of APIs in Sludge

The approach presented in this paper to quantify APIs in sludge is based on the comprehensive analytical methodology described by the United States Environmental Protection Agency; EPA Method1694 (U.S. Environmental Protection Agency 2007). Isotope dilution for calibration of each native compound was used when a labelled analogue was available, and calibration by internal standard was used to determine the concentration of the native compounds when no labelled compound was available. The two approaches have common mathematical operations. For the compounds determined by isotope dilution, the relative response (*RR*) (labelled to native) vs. concentration in the calibration solutions was computed over the calibration range according to Equation (2):

$$RR = (A_n C_l)/(A_l C_n) \tag{2}$$

A_n = the area of the daughter m/z for the native compound. A_l = the area of the daughter m/z for the labelled compound. C_l = the amount of the labelled compound in the calibration standard (pg). C_n = the amount of the native compound in the calibration standard (pg).

Response factors (*RF*) were calculated in a similar way; replacing A_l with A_{is} (area of the daughter m/z for the internal standard), C_l with C_{is} (amount of the internal standard, pg) and A_l with A_{is} (area of the daughter m/z for the internal standard).

The concentration of a native compound was calculated according to Equation (3):

$$C_n = (A_n C_l)/(A_l RR) \tag{3}$$

The calculated API concentration, C_n, was then used to express the content of an API in sludge as μg/kg dry weight sludge.

2.6.3. Statistical Analysis

The standard deviation was calculated for both the recovery of the 7 isotope labelled compounds in the internal standard mixture and the recovery for all 23 compounds investigated. ANOVA at a 95% confidence level were used to evaluate and compare the extraction efficiency of the *Espresso method* vs. the *PHWE method*.

3. Results and Discussion

3.1. Absolute Recoveries of Spiked Isotope Labelled Compounds and Ion Suppression

The eight isotope labelled compounds used in this study include positive, negative and neutral APIs, Table 1. Seven of these were spiked at the top of the extraction cell (internal standard mixture, Table S1), containing the sewage sludge, and underwent the entire extraction process, while the internal standard thiacloprid-D4 was spiked just prior to the analysis. By calculating the absolute recovery of the 7 labelled compounds the extraction efficiency of the Espresso method compared to *PHWE method* could be revealed. The obtained extraction efficiency results also contain a component of ion suppression, which might depend on the applied extraction methodology. To get basic information about possible differences in the magnitude of ion suppression caused by the two individual extraction methods the results for thiacloprid-D4 was used.

Table 1. Absolute recovery and relative standard deviation (RSD %, $n = 3$) for the instrumental standard *thiacloprid-d4* spiked just prior to the analysis and seven labelled standards spiked at the top of the two types of extraction cells, Espresso and PHWE, containing the sewage sludge. Absolute recovery was calculated using Equation (1). Four of the compounds were analysed at both basic and acid conditions and are marked in *italic*.

Compound	UPLC Condition	Espresso ($n = 3$) (%)	RSD (%)	PHWE ($n = 3$) (%)	RSD (%)
Thiacloprid-d4[a]	*Basic*	70	17	34	17
Carbamazepine-D10	*Basic*	44	18	28	4
Diclofenac-13C6	*Basic*	30	18	14	24
Sulfamethoxazole-13C6	*Basic*	53	7	26	19
Atenolol-d7	Basic	56	5	45	10
Methiocarb-d3	Basic	20	12	17	5
Metoprolol-d7	Basic	36	15	34	9
Estrone-d4	Basic	21	29	17	2
Average recovery 7 compounds	Basic	37		24	
Thiacloprid-d4[a]	*Acid*	37	22	14	18
Carbamazepine-d10	*Acid*	19	31	10	7
Diclofenac-13C6	*Acid*	20	11	14	12
Sulfamethoxazole-13C6	*Acid*	34	18	20	13
Average recovery 3 compounds	Acid	24		16	

The absolute recoveries obtained for all 8 compounds are shown in Table 1. It should be noted that four of these compounds were analysed in both the *acid* and the *basic chromatographic UPLC method*, giving information on possible differences in reduction of ion suppression by better separation of compounds from co-eluting matrix components during the two different UPLC runs. To provide additional knowledge of differences in terms of ion suppression, TIC spectra for both chromatographic methods and both extraction methods were also collected as shown in Figure 2.

Starting with the standard, thiacloprid-D4, added post extraction, it was shown that the absolute recovery for this compound was 70% in the Espresso method and 34% in the PHWE method with basic chromatographic conditions. This difference can be explained by a heavier matrix suppression expressed in the PHWE method, which is further supported by the higher TIC numbers for PHWE compared to Espresso, Figure 2a,b. Interestingly, there are also differences in absolute recovery comparing the thiacloprid-D4 results from the basic and acid chromatographic method separately. Recovery results dropped from the above-mentioned 70% in the basic method to 37% in the acid method, for samples run in the Espresso method. In the acid method, the recovery dropped from 34 to 14%. One plausible explanation is less ion suppression in the basic method. TIC spectra, generated by the PHWE method and the Espresso method, in the acid method is higher throughout the whole spectrum, reaching a maximum at 1.70 min and 1.6×10^{11} counts, compared to the basic method where the maximum is 1.2×10^{11} counts at 1.75 min, Figure 2a,b.

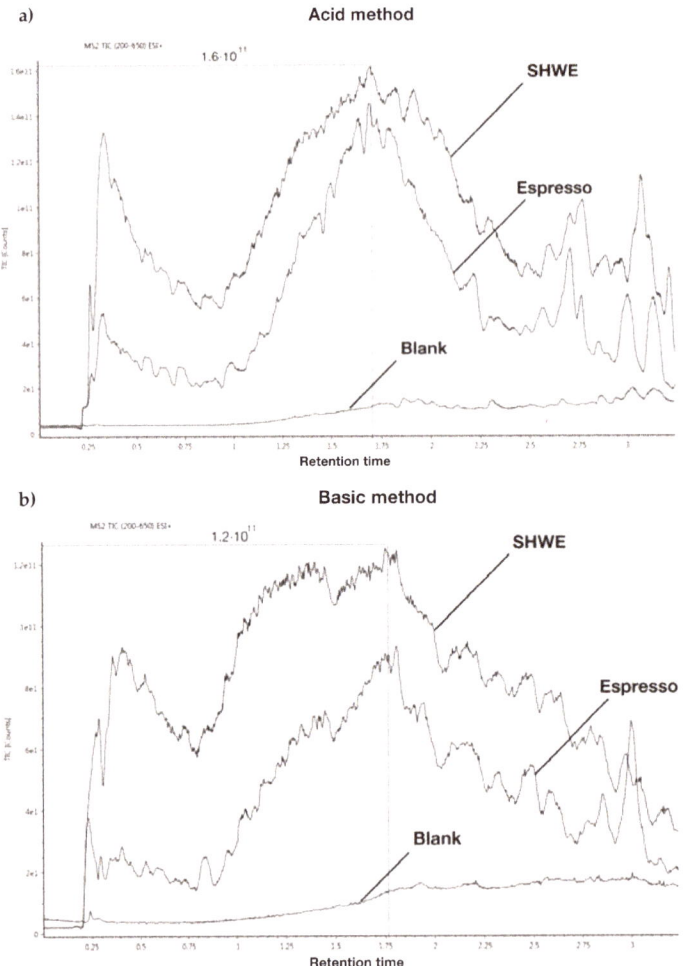

Figure 2. TIC spectra obtained using Waters RADAR^{TM} function of full scan (MS). Panel (**a**) shows TICs obtained for a blank sample as well as sewage sludge extracts applying the *Espresso method* and the *PHWE method* at acid chromatographic conditions. Panel (**b**) shows TICs for the same samples types but analyzed at basic chromatographic conditions.

From Figure 2a,b it is clear that the PHWE method extracts more sample components. Both above mentioned techniques rely upon the same shift in chemical and physical properties that water undergoes as the temperature is elevated. An increase in water temperature changes the relative permittivity from nearly $\varepsilon = 80$ at room temperature to $\varepsilon = 27$ at 250 °C. The organic solvents methanol and ethanol has a permittivity of $\varepsilon = 33$ and $\varepsilon = 24$ at 25 °C [36]. A declining degree of hydrogen bonding and decreased polarizablity can then also be observed. Thus, by increasing the water temperature the normal behaviour of water as a solvent will shift from polar to less-polar. Kondo and Yang compared the retention properties of superheated water and organic–water eluents and came to the conclusion that approximately 3.5 °C rise in water temperature was equivalent to 1% MeOH increase in MeOH–water mixtures [37]. This may be one reason for the higher co-extraction of matrix components using PHWE, since a 50 °C higher water temperature in PHWE would be equivalent to a 14% increase in MeOH content.

Turning to the three isotope labelled standards carbamazepine-D10, diclofenac-13C6 and sulfamethoxazole-13C6, which all were run in both the acid and the basic chromatographic methods, the effects of both the chromatographic conditions and the applied extraction method became even further accentuated, as seen in Table 1 and Figure 3a.

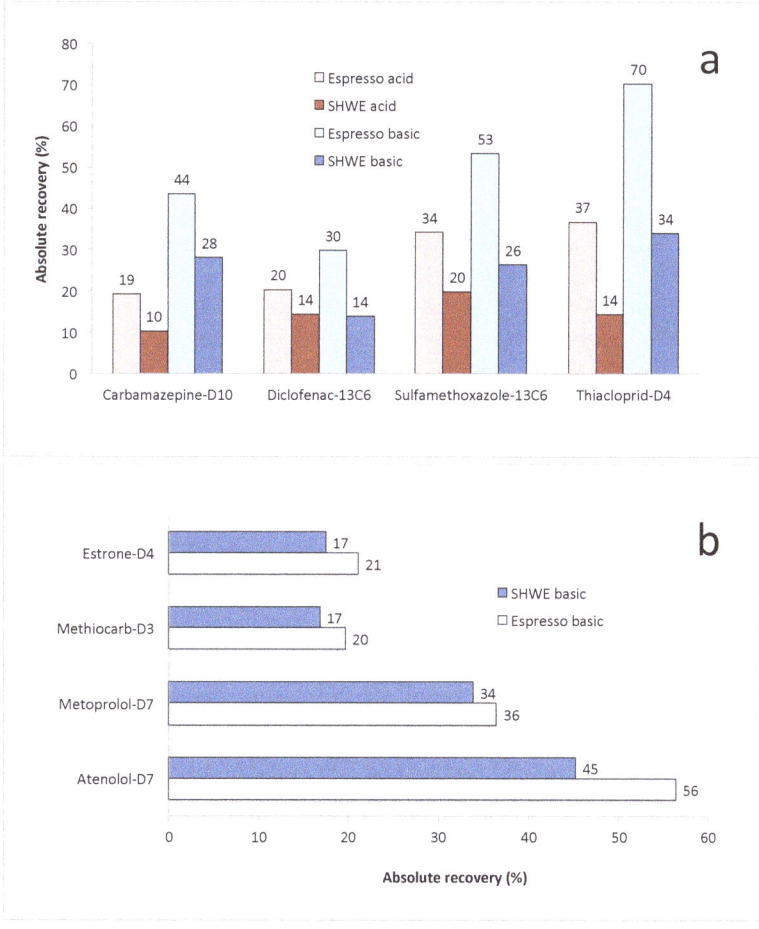

Figure 3. Absolute recovery for eight labelled standards spiked at the top of the extraction cell, containing the sewage sludge, and the instrumental standard thiacloprid-d4 spiked just prior to the analysis. The absolute recovery was calculated using Equation (1). Four analytes were analysed using two different chromatographic methods at acid and basic conditions (**a**), while four of the compounds could only be analysed at basic conditions (**b**).

Mean absolute recovery results for these labelled compounds in the basic method was 37% for the Espresso method and 24% for the PHWE method, while corresponding figures using the acid method were 24% and 16%, respectively. A first interpretation of the calculated mean values points in the direction of greater extraction efficiency using the Espresso method, but as discussed above, the results also reminds us of heavier ion suppression in the PHWE method. Göbel et al. [17] came to the conclusion that a temperature of 100 °C was optimal when optimizing pressurized liquid extraction (PLE) for the determination of sulfonamide, and macrolide antimicrobials and trimethoprim in sewage sludge. Increasing the temperature above 100 °C led to decreased recoveries. For sulfamethoxazole, they noticed a reduction by 95%, and for the macrolides the reduction was 60–90%. They concluded that this was due to thermal degradation of the analytes; however they also noted increasingly darker extracts at higher extraction temperatures, indicating a larger extraction of soluble organic matter. This notion of lower recoveries might not necessarily be a consequence of thermal degradation as we have shown in a temperature stability study [26], but rather a result of ion suppression, and which may partly be avoided as shown here by basic chromatography methods.

A second conclusion from Figure 3a would be that running the analysis under acid conditions always give lower absolute recoveries independent of compound and extraction method. This further strengthens the above statement made for thiacloprid-D4 that there may be less ion suppression in basic methods as is evident from TIC numbers shown in Figure 2. These results for isotope labelled standards clearly show the impact a chromatographic method might have when developing analytical methods for samples that express heavy ion suppression in mass spectrometry. Yet, a literature survey of methodologies published in high-impact analytical chemistry journals between 2006 and 2016 revealed that a vast majority of the identified methods rely on acidic mobile phases as discussed in a previous paper and thesis from our research group [35].

Among the four remaining isotope labelled compounds, atenolol-D7, methiocarb-D3, metoprolol-D7 and estrone-D4, there was a tendency that the PHWE method gave lower absolute recoveries as shown in Figure 3b, but not as pronounced as for the other compounds (Figure 3a). Atenolol-D7 showed the highest absolute recovery for both extraction methods, but also the largest difference between the two methods, with the Espresso method being most successful (56% recovery). Estrone-D4, methiocarb-D3 and metoprolol-D7 showed minor difference in recovery for the two methods.

3.2. Pharmaceuticals Extracted from Sludge

In a previous work dealing with surface water containing matrix components, we have determined which isotopic labelled standard that best compensates for losses caused by ion suppression and sample preparation losses for a specific API [34,35]. The pairing of labelled standard and API can be found in Table S2. The quantification in µg/kg sludge dry matter of a single API in the different sludge samples was calculated as described in Section 2.6.2. In total, 23 APIs were analysed. They have a molecular range from 236 to 734 D and contain 1-3 aromatic rings. A negative charge might be expressed from oxygenated substituents (e.g., diclofenac) and/or positive charge from nitrogen substituents (e.g., metoprolol). The results from the chemical analysis are presented in Table 2 together with various physicochemical parameters. Additionally, wastewater influent and effluent samples were collected at Kristianstad STP and the concentrations of all APIs were analysed and included in Table 2. Finally, results from a previous National Swedish Screening of API contents in sludge from the three Swedish cities Skövde, Stockholm and Umeå [38] are given for comparison at the end of Table 2 as are values from scientific articles on sludge published the last decade.

All 23 investigated compounds, except sulfamethoxazole, could be detected and quantified in sludge—with both extraction methods. In total, the content of APIs was 1042 µg/kg using the Espresso method and 5027 µg/kg using the PHWE method, i.e., nearly 5 times higher drug levels were estimated using PHWE. Average standard deviations were 22% for the Espresso method and 17% for the PHWE method. An ANOVA comparison of the two methods showed a significant difference ($F_{critical} = 7.7$) for 15 of the 23 investigated compounds and is marked in bold text in Table 2. One of the compounds, fluconazole, showed higher concentration with the Espresso method, while the remaining 14 compounds all were higher in the PHWE method. These differences can be explained, despite higher ion suppression as discussed earlier, by a more efficient extraction process using PHWE as it operates at a much higher temperature compared to Espresso. Nieto et al. 2010 [39] describes this in a clear way writing that: "Higher temperatures decrease the viscosity of liquid solvents, thus allowing better penetration of the matrix particles and enhancing extraction. In addition to reducing viscosity, high temperatures also decrease the surface tension of the solvent, the solutes and the matrix, allowing the solvent to "wet" the sample matrix more thoroughly". It has also been concluded that the time of operation have a strong influence on extraction efficiency, as stated by Nieto et al.: "The long exposure to the solvent allows the matrix to swell, thus improving penetration of solvent into the sample interstices and contact between solvent and analyte" [39]. In our study, the PHWE method operates under a longer time period, 13 min, as compared to the Espresso method extracting only for 10 s. As a consequence, more API molecules are likely liberated from the sewage sludge.

Turning to individual compounds, *five* compounds stands out; ciprofloxacin, citalopram, ketoconazole, metoprolol and venlafaxine, which all were detected in concentrations higher than 180 μg/kg using the *PHWE method*, all other APIs being below 70 μg/kg (Table 2, Figure 4). We would here like to make the reader aware of the logarithmic scale applied in Figure 4 due to the concentration range of APIs spanning three orders of magnitude

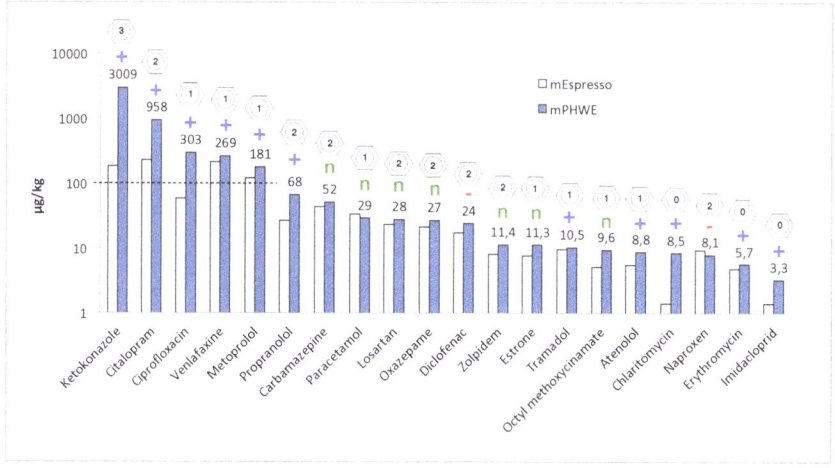

Figure 4. Concentrations in μg/kg of 23 APIs in lyophilized sewage sludge from Kristianstad STP determined using the *Espresso method* (light blue) and the *PHWE method* (dark blue). The APIs are listed in descending order of concentration (*PHWE method*) together with the charge and the number of aromatic ring structures of the compound in question.

The concentration found of ketoconazole in sludge using the *PHWE method* was 3009 μg/kg, which was 15 times higher as compared to the results from the *Espresso method*. Corresponding relative comparison between the methods for ciprofloxacin, citalopram, metoprolol and venlafaxine showed 5.1, 4.1, 1.5 and 1.2 times higher for the four compounds, respectively. In a previous investigation we showed that the matrix content of organic matter measured as total organic carbon content in sediments and sludge dictated drug sorption together with the presence of charged sites [40]. All of these top five compounds carry a positive charge at pH 7 (Table 2), and one of the explanations to these large amounts is probably the electrostatic interaction by the compounds positively charged groups and the negatively charged surfaces of the sludge matrix. Another part of the explanation is likely due to the existence of plural aromatic structures. Ketoconazole, which is by far the most abundant API found in the investigated sewage sludge samples contains three aromatic rings, citalopram, the second most abundant API contains two aromatic rings.

Table 2. Physicochemical properties, and determined concentrations in μg/kg of 23 APIs in lyophilized sewage sludge from Kristianstad STP using an *Espresso method* and a *PHWE method* (RSD %, *n* = 3). The methods were compared by ANOVA, and bold figures indicate statistically higher concentrations for that method. Inlet at outlet concentrations of the 23 APIs from Kristianstad STP are shown in the table along with previously determined API concentrations from 3 cities (Skövde, Stockholm and Umeå) in a Swedish National Screening Programme 2010. Literature data from scientific international studies are also presented at the end of the table.

Substance	pKa	Charge (pH 7)	Log D (pH 7)	log Kow	Espresso (*n* = 3) μg/kg	RSD (%)	PHWE (*n* = 3) μg/kg	RSD (%)	Water Inlet ng/L	Water Outlet ng/L	Skövde Sludge μg/kg	Stockholm Sludge μg/kg	Umeå Sludge μg/kg	Scientific Paper μg/kg	[REF]
Atenolol	9.67	+	-2.14	0.16	5.6	15	**8.8**	3	1348	215	13	9	12	6;22	[14]
Carbamazepine	-	n	2.28	2.45	43	26	52	10	1031	547	190	200	120	18;32	[11]
Clarithromycin	8.99	+	1.84	3.16	1.4	34	**8.5**	22	131	22	13	1.4	4.5	24	[17]
Ciprofloxacin	6.38	(+;-)	-0.81	0.28	60	38	**303**	40	58	46	450	250	170	2420	[12]
Citalopram	9.78	+	1.27	2.51	233	31	**958**	8	155	32	760	570	630	725	[13]
Diclofenac	4.00	-	1.37	4.06	18	16	24	11	713	577	59	31	10	192	[8]
Erythromycin	8.80	+	1.2	2.50	4.8	30	5.7	8	385	267	1000	150	120	62	[8]
Estrone	10.30	n	4.31	3.13	7.9	6	**11**	9	49	3	33	36	2	23;28	[14]
Fluconazole	2.56	n	0.56	0.25	**0.5**	15	0.4	12	51	105	3.5	13	47	<LOQ; n.d	[1]
Imidacloprid	5.28	pH6 +	1.09	0.57	1.4	38	**3.3**	24	8	14	n.a	n.a	n.a		
Ketokonazole	3.96, 6.75	+	4.06	4.30	186	36	**3009**	12	69	0	510	1200	1100	910	[15]
Losartan	4.12	n	4.94	4.01	23	7	28	22	326	205	n.a	n.a	n.a		
Metoprolol	9.60	+	-0.81	1.88	123	21	**181**	7	999	533	180	410	210	29.92	[16]
Naproxen	4.19	-	0.45	3.18	9.3	25	8.1	13	2421	290	<LOQ	<LOQ	<LOQ	4	[17]
Octylmethoxycinamate	-	n	5.38	6.10	5.2	5	**10**	31	64	31	n.a	n.a	n.a		
Oxazepame	-	n	2.06	2.31	22	33	27	14	374	403	43	18	12		
Paracetamol	9.46	n	0.74	1.08	34	29	29	33	22528	0	73	11	<LOQ		
Propranolol	9.42	+	1.15	3.48	27	22	**68**	9	47	16	n.a	n.a	n.a	<1.26	[8]
Sulfamethoxazole	6.16	-	0.14	0.89	n.d		n.d		476	118	<LOQ	<LOQ	<LOQ	n.d	[8]
Tramadol	9.23	+	0.24	2.51	10	12	11	10	168	153	<LOQ	68	<LOQ	43	[19]
Trimethoprim	7.20	+	0.92	0.91	0.2	33	**1.2**	19	77	17	27	2.2	2.5	5;13	[8]
Venlafaxine	8.91	+	0.84	3.28	219	7	**269**	20	174	88	86	310	150	318	[1]
Zolpidem	5.65	n	3	3.85	8.4	9	**11**	26	12	4	7.7	8.3	3.2	38	[19]
Sum concentration					1042		5027		31664	3686	3448	3288	2593		

From Table 2 it can be seen that of the 23 compounds analysed, 20 showed a concentration in sludge exceeding 1 µg/kg sludge. These are all shown visually in descending order of concentration in Figure 4, and reveals that 18 of these compounds are all in favour of the *PHWE method*. Figure 4 also shows the clear pattern of positively charged molecules being at the highest concentrations range in the sludge as discussed above, followed by neutrals, and some additional positively and negatively charged compounds. The reason for finding some of the positively charged compounds on the far-right side could likely be attributed to two circumstances. The most obvious is that some of these compounds simply do not occur at high enough concentrations in the incoming water to reach high concentrations in the sludge despite a high degree of affinity to this negatively charged matrix. This is further discussed in the next section presenting APIs in the incoming wastewater. The other option is, as already touched upon, is that there are other chemical interactions at play. One of these is the aromatic ring structure. As mentioned, the two top compounds, ketoconazole and citalopram, have three and two aromatic rings. To reveal any possible relation to aromaticity, Figure 4 also presents the number of aromatic rings present on each of the 20 APIs. From this it becomes evident that those neutral compounds that reach fairly high concentrations in most cases have 2 aromatic rings, the one exception being paracetamol. Yet, finding paracetamol at such high concentrations in sludge might not come as too much of a surprise considering the large amounts of this compound being sold over the counter in Sweden, and being found in enormous concentrations in incoming water (see below). Additionally, on the very far right end there are 4 positively charged APIs. Three of these have no aromatic ring structure at all. In conclusion, there are strong indications that both the number of aromatic rings and the presence of a positive charge are important for the sorption processes of APIs to sewage sludge.

3.3. Comparison of Extraction Results with Effluent and Influent Data

In Table 2 we have included wastewater influent and effluent analysis results for the investigated APIs from Kristianstad STP, performed by us according to our previously published methodology [34,35]. The first observation was that all 23 APIs were present in the *influent* water. Likewise, all investigated APIs except sulfamethoxazole were present in the sewage sludge samples. In the effluent samples, all APIs except ketoconazole and paracetamol could be detected. Based on results from Figure 4 and Table 2 it is likely that the efficient reduction of ketoconazole in the STP primarily is caused by sludge adsorption. Paracetamol on the other hand show much lower sludge concentrations, and the huge loss of paracetamol, with the highest influent values, must be explained by either abiotic or biotic degradation/transformation, or a combination thereof.

Evidently the sludge concentrations measured in PHWE and presented together with two physicochemical properties in Figure 4, and inlet and outlet concentrations cannot fully explain the observed pattern. Yet, a few more chemical remarks can be made in favour of using charge and aromaticity as valuable additional tools in describing sludge sorption apart from primarily looking at more well-established classical parameters such as log K_{ow} and log D.

For example, in Table 2 high log K_{ow} and log D values can be found for ketoconazole, but not for ciprofloxacin and citalopram. Here charge may aid in better explaining the high concentration of the two latter on sludge. Metoprolol and carbamazepine have approximately the same influent and effluent concentrations, yet metoprolol is 3.5 times more abundant in sludge compared to carbamazepine, despite that log K_{ow} and log D is higher for carbamazepine than metoprolol. This again shows that the positive charge might have a strong impact on sorption. These results show that neither log K_{ow} nor log D alone are valid tools too fully describing APIs adsorption to sludge.

3.4. Comparison of Extraction Results with Literature Data

In Table 2 we have also included analysis results from the Swedish National Screening Programme 2010 [38] together with a selection of literature data describing concentrations of APIs in digested sludge. The top five compounds in the Swedish National Screening Programme were ketoconazole, citalopram, ciprofloxacin, metoprolol, and venlafaxine, the same top five compounds that we found in our extracted sludge. With a few exceptions, the absolute values are roughly the same, only the antibiotic compound erythromycin showed a lower concentration in our study compared to the National Screening Programme, representing the only truly deviating value.

In the Swedish report, we couldn't find what extraction technique and method used. If not as exhaustive as our PHWE method this may be one reason for not achieving such high sum concentrations of all investigated compounds. In fact, the two compounds contributing most to the difference is ketoconazole and citalopram, both adsorbing strongly to the sludge. In our work ketoconazole had a concentration of 3009 µg/kg, while the other three sludges (Skövde, Stockholm and Umeå) were in the range 510–1200 µg/kg. Likewise, citalopram had a concentration in our study of 958 µg/kg in our study but was in the range 570–760 µg/kg in the Swedish report.

4. Conclusions

Two green chemistry extraction systems based on either an in-house stainless-steel column pressurized hot water extraction system (PHWE method) or a commercially available Espresso machine (Espresso method) were applied for analysing 23 APIs in sewage sludge. Both methods showed god repeatability. The PHWE method allowed a more complete extraction of APIs that were more tightly bound to the matrix, as exemplified by the results from ketoconazole, citalopram and ciprofloxacin. In fact, 19 out of 23 investigated APIs were quantified in sludge, and with few exceptions they showed higher concentrations using the PHWE method. The Espresso method might still be useful if strong electrostatic forces aren't to be overcome as it gives cleaner extracts (lower ion suppression) and is much faster to operate compared to the PHWE method, allowing higher sample throughput. An Espresso extraction takes 10 s, while the PHWE extraction takes several minutes. The results produced with the PHWE method are in line with results reported from a Swedish National screening program and international literature data. A weakness with both methods, which was more pronounced for the PHWE method, were the low absolute recoveries, which most likely were caused by matrix ion suppression. Further investigations will focus on reducing these effects by investigating fractionating of the extracts and the possibility of matrix precipitation using various techniques. Additionally, UPLC-MS/MS, applying basic conditions, is surprisingly unexplored in environmental analysis, despite its positive effects in reducing ion suppression, and should be further investigated.

Supplementary Materials: The following are available online at http://www.mdpi.com/2076-3417/9/7/1509/s1, Table S1: Compounds analysed, Table S2: Labelled standard used for each API.

Author Contributions: Conceptualization, O.S.; methodology, O.S.; validation, O.S. and E.B.; formal analysis, O.S.; software, O.S.; investigation, O.S.; resources, O.S.; data curation, O.S.; writing—original draft preparation, O.S.; writing—review and editing, O.S. and E.B.; visualization, O.S. and E.B.; supervision, O.S.; project administration, O.S.; funding acquisition, O.S. and E.B.

Funding: This research received no external funding

Conflicts of Interest: The authors declare no conflict of interest.

References

1. Wahlberg, C.; Björlenius, B.; Ek, M.; Paxéus, N.; Gårdstam, L. *Avloppsreningsverkens Förmåga att ta Hand om Läkemedelsrester och Andra farliga Ämnen*; Naturvårdsverket: Stockholm, Sweden, 2008; pp. 1–162.
2. Halling-Sørensen, B.; Nielsen, S.N.; Lanzky, P.F.; Ingerslev, F.; Lützhøft, H.H.; Jørgensen, S. Occurrence, fate and effects of pharmaceutical substances in the environment-A review. *Chemosphere* **1998**, *36*, 357–393. [CrossRef]
3. Pérez-Carrera, E.; Hansen, M.; León, V.M.; Björklund, E.; Krogh, K.A.; Halling-Sørensen, B.; González-Mazo, E. Multiresidue method for the determination of 32 human and veterinary pharmaceuticals in soil and sediment by pressurized-liquid extraction and LC-MS/MS. *Anal. Bioanal. Chem.* **2010**, *398*, 1173–1184. [CrossRef] [PubMed]

4.   Rodríguez-Navas, C.; Björklund, E.; Bak, S.A.; Hansen, M.; Krogh, K.A.; Maya, F.; Cerdà, V. Pollution pathways of pharmaceutical residues in the aquatic environment on the island of Mallorca, Spain. *Arch. Environ. Contam. Toxicol.* **2013**, *65*, 56–66. [CrossRef] [PubMed]

5.   Sui, Q.; Cao, X.; Lu, S.; Zhao, W.; Qiu, Z.; Yu, G. Occurrence, sources and fate of pharmaceuticals and personal care products in the groundwater: A review. *Emerg. Contam.* **2015**, *1*, 14–24. [CrossRef]

6.   Osorio, V.; Larrañaga, A.; Aceña, J.; Pérez, S.; Barceló, D. Concentration and risk of pharmaceuticals in freshwater systems are related to the population density and the livestock units in Iberian Rivers. *Sci. Total Environ.* **2016**, *540*, 267–277. [CrossRef] [PubMed]

7.   Paíga, P.; Santos, L.H.; Ramos, S.; Jorge, S.; Silva, J.G.; Delerue-Matos, C. Presence of pharmaceuticals in the Lis river (Portugal): Sources, fate and seasonal variation. *Sci. Total Environ.* **2016**, *573*, 164–177. [CrossRef] [PubMed]

8.   Kay, P.; Hughes, S.R.; Ault, J.R.; Ashcroft, A.E.; Brown, L.E. Widespread, routine occurrence of pharmaceuticals in sewage effluent, combined sewer overflows and receiving waters. *Environ. Pollut.* **2017**, *220*, 1447–1455. [CrossRef]

9.   Björklund, E.; Svahn, O.; Bak, S.; Bekoe, S.O.; Hansen, M. Pharmaceutical residues affecting the UNESCO biosphere reserve Kristianstads Vattenrike wetlands: Sources and sinks. *Arch. Environ. Contam. Toxicol.* **2016**, *71*, 423–436. [CrossRef]

10.   Verlicchi, P.; Zambello, E. Pharmaceuticals and personal care products in untreated and treated sewage sludge: Occurrence and environmental risk in the case of application on soil—A critical review. *Sci. Total Environ.* **2015**, *538*, 750–767. [CrossRef]

11.   Jederlund, L. Frågor och svar, REVAQ, 160108. 2016.

12.   Hansson, E.; Johansson, M. *Rapport Avlopp på våra åkrar—En Rapport om Miljögifter i slam*; Naturskyddsföreningen: Stockholm, Sweden, 2012; pp. 1–26.

13.   Petrie, B.; Youdan, J.; Barden, R.; Kasprzyk-Hordern, B. Multi-residue analysis of 90 emerging contaminants in liquid and solid environmental matrices by ultra-high-performance liquid chromatography tandem mass spectrometry. *J. Chromatogr. A* **2016**, *1431*, 64–78. [CrossRef]

14.   Barron, L.; Tobin, J.; Paull, B. Multi-residue determination of pharmaceuticals in sludge and sludge enriched soils using pressurized liquid extraction, solid phase extraction and liquid chromatography with tandem mass spectrometry. *J. Environ. Monit.* **2008**, *10*, 353–361. [CrossRef] [PubMed]

15.   Ekpeghere, K.I.; Lee, J.W.; Kim, H.Y.; Shin, S.K.; Oh, J.E. Determination and characterization of pharmaceuticals in sludge from municipal and livestock wastewater treatment plants. *Chemosphere* **2017**, *168*, 1211–1221. [CrossRef] [PubMed]

16.   Lindberg, R.H.; Olofsson, U.; Rendahl, P.; Johansson, M.I.; Tysklind, M.; Andersson, B.A. Behavior of fluoroquinolones and trimethoprim during mechanical, chemical, and active sludge treatment of sewage water and digestion of sludge. *Environ. Sci. Technol.* **2006**, *40*, 1042–1048. [CrossRef] [PubMed]

17.   Göbel, A.; Thomsen, A.; McArdell, C.S.; Alder, A.C.; Giger, W.; Theiß, N.; Ternes, T.A. Extraction and determination of sulfonamides, macrolides, and trimethoprim in sewage sludge. *J. Chromatogr. A* **2005**, *1085*, 179–189. [CrossRef]

18.   Radjenović, J.; Jelić, A.; Petrović, M.; Barceló, D. Determination of pharmaceuticals in sewage sludge by pressurized liquid extraction (PLE) coupled to liquid chromatography-tandem mass spectrometry (LC-MS/MS). *Anal. Bioanal. Chem.* **2009**, *393*, 1685–1695. [CrossRef] [PubMed]

19.   Runnqvist, H.; Bak, S.A.; Hansen, M.; Styrishave, B.; Halling-Sørensen, B.; Björklund, E. Determination of pharmaceuticals in environmental and biological matrices using pressurised liquid extraction—Are we developing sound extraction methods? *J. Chromatogr. A* **2010**, *1217*, 2447–2470. [CrossRef] [PubMed]

20.   Luque-Munoz, A.; Vílchez, J.L.; Zafra-Gómez, A. Multiclass method for the determination of pharmaceuticals and personal care products in compost from sewage sludge using ultrasound and salt-assisted liquid–liquid extraction followed by ultrahigh performance liquid chromatography-tandem mass spectrometry analysis. *J. Chromatogr. A* **2017**, *1507*, 72–83. [CrossRef]

21.   Clark, J.H. Green chemistry: Challenges and opportunities. *Green Chem.* **1999**, *1*, 1–8. [CrossRef]

22.   Anastas, P.; Eghbali, N. Green chemistry: Principles and practice. *Chem. Soc. Rev.* **2010**, *39*, 301–312. [CrossRef]

23.   Kronholm, J.; Hartonen, K.; Riekkola, M.L. Analytical extractions with water at elevated temperatures and pressures. *Trac Trends Anal. Chem.* **2007**, *26*, 396–412. [CrossRef]

24. Hawthorne, S.B.; Yang, Y.; Miller, D.J. Extraction of organic pollutants from environmental solids with sub-and supercritical water. *Anal. Chem.* **1994**, *66*, 2912–2920. [CrossRef]

25. Stoob, K.; Singer, H.P.; Stettler, S.; Hartmann, N.; Mueller, S.R.; Stamm, C.H. Exhaustive extraction of sulfonamide antibiotics from aged agricultural soils using pressurized liquid extraction. *J. Chromatogr. A* **2006**, *1128*, 1–9. [CrossRef]

26. Svahn, O.; Björklund, E. Thermal stability assessment of antibiotics in moderate temperature and subcriticalwater using a pressurized dynamic flow-through system. *Int. J. Innov. Appl. Stud.* **2015**, *11*, 872–880.

27. Sánchez-López, J.A.; Zimmermann, R.; Yeretzian, C. Insight into the time-resolved extraction of aroma compounds during espresso coffee preparation: Online monitoring by PTR-ToF-MS. *Anal. Chem.* **2014**, *86*, 11696–11704. [CrossRef]

28. Mestdagh, F.; Davidek, T.; Chaumonteuil, M.; Folmer, B.; Blank, I. The kinetics of coffee aroma extraction. *Food Res. Int.* **2014**, *63*, 271–274. [CrossRef]

29. Severini, C.; Ricci, I.; Marone, M.; Derossi, A.; De Pilli, T. Changes in the aromatic profile of espresso coffee as a function of the grinding grade and extraction time: A study by the electronic nose system. *J. Agric. Food Chem.* **2015**, *63*, 2321–2327. [CrossRef]

30. Salamanca, C.A.; Fiol, N.; González, C.; Saez, M.; Villaescusa, I. Extraction of espresso coffee by using gradient of temperature. Effect on physicochemical and sensorial characteristics of espresso. *Food Chem.* **2017**, *214*, 622–630. [CrossRef]

31. Armenta, S.; de la Guardia, M.; Esteve-Turrillas, F.A. Hard cap espresso machines in analytical chemistry: What else? *Anal. Chem.* **2016**, *88*, 6570–6576. [CrossRef] [PubMed]

32. López, A.; Coscollà, C.; Yusà, V.; Armenta, S.; de la Guardia, M.; Esteve-Turrillas, F.A. Comprehensive analysis of airborne pesticides using hard cap espresso extraction-liquid chromatography-high-resolution mass spectrometry. *J. Chromatogr. A* **2017**, *1506*, 27–36.

33. Gallart-Mateu, D.; Pastor, A.; de la Guardia, M.; Armenta, S.; Esteve-Turrillas, F.A. Hard cap espresso extraction-stir bar preconcentration of polychlorinated biphenyls in soil and sediments. *Anal. Chim. Acta* **2017**, *952*, 41–49. [CrossRef] [PubMed]

34. Svahn, O. Tillämpad Miljöanalytisk kemi för Monitorering och Åtgärder av Antibiotika-och Läkemedelsrester i Vattenriket. Doctoral Dissertation, Lund University, Lund, Sweden, 2016.

35. Svahn, O.; Björklund, E. Increased electrospray ionization intensities and expanded chromatographic possibilities for emerging contaminants using mobile phases of different pH. *J. Chromatogr. B* **2016**, *1033*, 128–137. [CrossRef]

36. Teo, C.C.; Tan, S.N.; Yong, J.W.H.; Hew, C.S.; Ong, E.S. Pressurized hot water extraction (PHWE). *J. Chromatogr. A* **2010**, *1217*, 2484–2494. [CrossRef]

37. Kondo, T.; Yang, Y. Comparison of elution strength, column efficiency, and peak symmetry in subcritical water chromatography and traditional reversed-phase liquid chromatography. *Anal. Chim. Acta* **2003**, *494*, 157–166. [CrossRef]

38. Fick, J.; Lindberg, R.H.; Brorström-Lundén, E. *Results from the Swedish National Screening Programme 2010*; IVL, Swedish Environmental Research Institute Ltd.: Stockholm, Sweden, 2011; pp. 1–56.

39. Nieto, A.; Borrull, F.; Pocurull, E.; Marcé, R.M. Pressurized liquid extraction: A useful technique to extract pharmaceuticals and personal-care products from sewage sludge. *Trac Trends Anal. Chem.* **2010**, *29*, 752–764. [CrossRef]

40. Svahn, O.; Björklund, E. Describing sorption of pharmaceuticals to lake and river sediments, and sewage sludge from UNESCO Biosphere Reserve Kristianstads Vattenrike by chromatographic asymmetry factors and recovery measurements. *J. Chromatogr. A* **2015**, *1415*, 73–82. [CrossRef] [PubMed]

41. da Silva, B.F.; Jelic, A.; López-Serna, R.; Mozeto, A.A.; Petrovic, M.; Barceló, D. Occurrence and distribution of pharmaceuticals in surface water, suspended solids and sediments of the Ebro river basin, Spain. *Chemosphere* **2011**, *85*, 1331–1339. [CrossRef] [PubMed]

42. Golet, E.M.; Strehler, A.; Alder, A.C.; Giger, W. Determination of fluoroquinolone antibacterial agents in sewage sludge and sludge-treated soil using accelerated solvent extraction followed by solid-phase extraction. *Anal. Chem.* **2002**, *74*, 5455–5462. [CrossRef] [PubMed]

43. Lajeunesse, A.; Smyth, S.A.; Barclay, K.; Sauvé, S.; Gagnon, C. Distribution of antidepressant residues in wastewater and biosolids following different treatment processes by municipal wastewater treatment plants in Canada. *Water Res.* **2012**, *46*, 5600–5612. [CrossRef] [PubMed]

44. Andersen, H.R.; Hansen, M.; Kjølholt, J.; Stuer-Lauridsen, F.; Ternes, T.; Halling-Sørensen, B. Assessment of the importance of sorption for steroid estrogens removal during activated sludge treatment. *Chemosphere* **2005**, *61*, 139–146. [CrossRef] [PubMed]

45. Östman, M.; Lindberg, R.H.; Fick, J.; Björn, E.; Tysklind, M. Screening of biocides, metals and antibiotics in Swedish sewage sludge and wastewater. *Water Res.* **2017**, *115*, 318–328. [CrossRef]

46. Ramil, M.; El Aref, T.; Fink, G.; Scheurer, M.; Ternes, T.A. Fate of beta blockers in aquatic-sediment systems: Sorption and biotransformation. *Environ. Sci. Technol.* **2009**, *44*, 962–970. [CrossRef] [PubMed]

47. Malmborg, J.; Magnér, J. Pharmaceutical residues in sewage sludge: Effect of sanitization and anaerobic digestion. *J. Environ. Manag.* **2015**, *153*, 1–10. [CrossRef] [PubMed]

48. Martín, J.; Camacho-Muñoz, D.; Santos, J.L.; Aparicio, I.; Alonso, E. Occurrence of pharmaceutical compounds in wastewater and sludge from wastewater treatment plants: Removal and ecotoxicological impact of wastewater discharges and sludge disposal. *J. Hazard. Mater.* **2012**, *239*, 40–47. [CrossRef] [PubMed]

49. Peysson, W.; Vulliet, E. Determination of 136 pharmaceuticals and hormones in sewage sludge using quick, easy, cheap, effective, rugged and safe extraction followed by analysis with liquid chromatography–time-of-flight-mass spectrometry. *J. Chromatogr. A* **2013**, *1290*, 46–61. [CrossRef] [PubMed]

Article

A Novel Method to Characterise Levels of Pharmaceutical Pollution in Large-Scale Aquatic Monitoring Campaigns

John L. Wilkinson [1,*], **Alistair B.A. Boxall** [1] and **Dana W. Kolpin** [2]

[1] Environment and Geography Department, University of York, York YO10 5NG, UK;
  alistair.boxall@york.ac.uk
[2] U.S. Geological Survey, Central Midwest Water Science Center, Iowa City, IA 52240, USA;
  dwkolpin@usgs.gov
* Correspondence: john.wilkinson@york.ac.uk

Received: 11 February 2019; Accepted: 20 March 2019; Published: 1 April 2019

Abstract: Much of the current understanding of pharmaceutical pollution in the aquatic environment is based on research conducted in Europe, North America and other select high-income nations. One reason for this geographic disparity of data globally is the high cost and analytical intensity of the research, limiting accessibility to necessary equipment. To reduce the impact of such disparities, we present a novel method to support large-scale monitoring campaigns of pharmaceuticals at different geographical scales. The approach employs the use of a miniaturised sampling and shipping approach with a high throughput and fully validated direct-injection High-Performance Liquid Chromatography-Tandem Mass Spectrometry method for the quantification of 61 active pharmaceutical ingredients (APIs) and their metabolites in tap, surface, wastewater treatment plant (WWTP) influent and WWTP effluent water collected globally. A 7-day simulated shipping and sample stability assessment was undertaken demonstrating no significant degradation over the 1–3 days which is typical for global express shipping. Linearity (r^2) was consistently ≥ 0.93 (median = 0.99 ± 0.02), relative standard deviation of intra- and inter-day repeatability and precision was <20% for 75% and 68% of the determinations made at three concentrations, respectively, and recovery from Liquid Chromatography Mass Spectrometry grade water, tap water, surface water and WWTP effluent were within an acceptable range of 60–130% for 87%, 76%, 77% and 63% of determination made at three concentrations respectively. Limits of detection and quantification were determined in all validated matrices and were consistently in the ng/L level needed for environmentally relevant API research. Independent validation of method results was obtained via an interlaboratory comparison of three surface-water samples and one WWTP effluent sample collected in North Liberty, Iowa (USA). Samples used for the interlaboratory validation were analysed at the University of York Centre of Excellence in Mass Spectrometry (York, UK) and the U.S. Geological Survey National Water Quality Laboratory in Denver (Colorado, USA). These results document the robustness of using this method on a global scale. Such application of this method would essentially eliminate the interlaboratory analytical variability typical of such large-scale datasets where multiple methods were used.

Keywords: pharmaceuticals; organic pollutants; liquid chromatography tandem mass spectrometry; validation; global monitoring

---

1. Introduction

Over the last 20 years, active pharmaceutical ingredients (APIs) and their metabolites have been identified in all environmental compartments and their occurrence has raised concerns over potential impacts on ecosystem and human health [1–3]. However, despite two decades of research, significant knowledge gaps exist regarding the environmental exposures to APIs [1,4]. For example, a complete or significant lack of knowledge exists for many parts of the world (e.g., Africa and South America) as such research disproportionally targets wealthy regions including North America, Western Europe and China [4,5]. In many poorly studied parts of the world, APIs are openly available without a prescription and prone to miss-/over-use so concentrations might be expected to be greater than those reported so far [6]. Similarly, API disposal practices and inefficient wastewater connectivity and treatment may further exacerbate high API concentrations in some regions [1,6]. Underpinning such research are the complex analytical methods employed for specific quantification of APIs in water, namely High-Performance Liquid Chromatography-Tandem Mass Spectrometry (HPLC-MS/MS). Such instrumentation is required for obtaining environmentally relevant sensitivity in measurements of APIs, down to ng/L levels.

The high cost of necessary instruments and analytical intensity of the methods employed are key barriers to broadening measurement of APIs in the environment to a global scale. Furthermore, such barriers may magnify existing regional disproportionalities in published data. Of the available published data, among the most significant challenges for compiling an international perspective on APIs in the aquatic environment is the cross-comparison between datasets obtained via different methodologies. Key areas of deviation between methodologies include sample collection protocols (e.g., collection from the river bank vs. centroid flow), analytical techniques and statistical/quantitative interpretations and use of quality-control samples throughout collection and analysis. No single, unified analytical method exists, and in-house method validations are not always required for publication, making accurate and reliable interpretations of existing data on concentrations of APIs in different regions of the globe challenging.

Recent advances in the sensitivity of analytical instrumentation provides the ability to now analyse a wide range of APIs with minimal sample pre-treatment (e.g., Furlong et al. [7]; Campos-Mañas et al. [8]. Direct-injection HPLC-MS/MS is characterised by large-volume sample injections (100–5000 μL) which eliminate the need for sample pre-concentration [9], traditionally achieved using solid phase extraction (SPE). This technique has been successfully used since the 1980s [9] and more recently employed for quantification of pharmaceuticals, illicit drugs and other organic contaminants in surface and wastewater [7,8,10]. Such methods significantly reduce the volume of sample needed (simplifying both sample collection and shipment to the laboratory) and the time for sample analysis via elimination of complex pre-treatment. Additionally, such methods offer a more environmentally responsible alternative to traditional methods (e.g., SPE) due to reduced sample volume and elimination of any need for solvents in sample pre-treatment. Direct-injection protocols also provide an opportunity to perform much larger-scale monitoring programmes than previously possible (e.g., due to financial and time constraints), allowing a better understanding of environmental exposures to pharmaceutical compounds (APIs and corresponding API metabolites) to aquatic systems around the globe. Conducting such large-scale global monitoring programmes will likely raise logistical challenges in terms of sample transport from the site of collection to the site of analysis. For these studies to succeed, therefore, it is important that sample integrity is maintained during such transport.

Here we present and evaluate a monitoring approach for use in large-scale monitoring programmes for APIs in multiple environmental matrices (e.g., WWTP influents, effluents, surface water and drinking water). The approach presents a simple and standardized set of protocols for the consistent collection, shipment, and analysis of aqueous samples using a uniform collection kit and a single HPLC-MS/MS analytical method. An interlaboratory evaluation of the protocol was conducted with surface water (i.e., river water) and WWTP effluent collected by the U.S. Geological Survey (USGS) from an effluent-impacted stream (Muddy Creek, North Liberty, Iowa, USA). This protocol may significantly reduce the challenges of: (a) evaluating spatial and temporal API concentration trends across variations in geography, climate, land use, hydrogeology, and demographics, (b) the lack of accessibility to costly analytical equipment and operating costs necessary for accurate and sensitive API quantification (namely HPLC-MS/MS) in some regions of the world, and (c) obtaining water samples from under-studied regions of the world for accurate, sensitive, reliable sample analysis. The development of this protocol therefore provides an opportunity to begin to better understand the risks of pharmaceuticals and other compounds at the global scale, a research priority highlighted in recent horizon scanning exercises on pharmaceuticals and chemicals more generally [5,11].

2. Materials and Methods

2.1. Test Substances

The protocol was developed for quantifying concentrations of 61 strategically selected APIs (Table 1) representing 19 therapeutic classes of medicinal chemicals approved for use in humans ($n = 57$) and animal husbandry ($n = 4$). The study APIs were selected to include: (a) compounds of high usage across the world; (b) compounds with known or suspected ecological or human health concern; and (c) compounds of expected high use due to regional disease pressures (e.g., antimalarials). Significant focus was placed on antimicrobial chemicals including antibiotics ($n = 13$) and antifungals ($n = 6$) due to implications for the selection of resistance to these medicines in bacterial communities [12,13]. Similarly, focus was also placed on antidepressants ($n = 7$) due to their increasing use globally and potential ecotoxicological risk [14]. All compounds were optimised for specific quantification using direct injection HPLC-MS/MS only and compounds not suitable for quantification using this instrumentation were not included in the study. Further method development may be used to broaden the scope of the studied contaminants (e.g., Campos-Mañas et al. [8]).

All test standards were purchased from Sigma Aldrich (UK) and were of ≥95% purity. Deuterated internal standards were obtained from Sigma Aldrich (UK) for 32 test APIs and atrazine-D5 was used where a labelled standard was not available. Liquid chromatography-mass spectrometry (LCMS)-grade water and methanol were obtained from VWR (UK). Polystyrene boxes (5 L, 34.5 × 21 × 14.5 cm, L × W × H) used for sample shipment were obtained from JB Packaging (Torpoint, UK), whereas 15-mL amber glass sample vials, 0.7-µm glass microfiber syringe filters (Whatman) and 24-mL luer lock syringes were obtained from Fisher Scientific (UK). A ZORBAX Eclipse Plus C18 chromatography column (3.0 × 100 mm, 1.8 µm, 600 bar) was purchased from Agilent Technologies (UK) and a C18 SecurityGuard guard column was purchased from Phenomenex (UK).

Table 1. List of monitored chemicals by therapeutic class and their associated internal standard.

Therapeutic Class		Compound	Associated Internal Standard
Analgesic		Lidocaine	Lidocaine D6
		Naproxen	Naproxen D3
		Paracetamol	Paracetamol D4
Anti-epileptic		Carbamazepine	Carbamazepine D10
		Gabapentin	Gabapentin D10
		Pregabalin	Atrazine D5
Antibiotics	*Fluoroquinolones*	Ciprofloxacin	Ciprofloxacin D8
		Enrofloxacin *	Atrazine D5
	Lincosamide	Lincomyacin *	Atrazine D5
	Macrolides	Clarithromycin	Atrazine D5
		Erythromycin	Atrazine D5
		Tylosin *	Atrazine D5
	Nitroimidazole	Metronidazole	Metronidazole D3
	Penicillin	Cloxacillin	Atrazine D5
	Sulfonamides	Sulfadiazine *	Atrazine D5
		Sulfamethoxazole	Sulfamethoxazole D4
		Trimethoprim	Trimethoprim D9
	Tetracyclines	Oxytetracycline	Atrazine D5
		Tetracycline	Atrazine D5

Table 1. *Cont.*

Therapeutic Class	Compound	Associated Internal Standard
Antidepressant	Amitriptyline Citalopram Desvenlafaxine Fluoxetine Sertraline Venlafaxine Norfluoxetine	Amitriptyline D3 Citalopram D6 Desvenlafaxine D6 Atrazine D5 Sertraline D3 Venlafaxine D6 Norfluoxetine D6
Antifungal	Clotrimazole Fluconazole Itraconazole Ketoconazole Miconazole Thiabendazole	Atrazine D5 Atrazine D5 Itraconazole D4 Atrazine D5 Atrazine D5 Atrazine D5
Antihistamine	Cetirizine Diphenhydramine Fexofenadine Ketotifen Loratadine	Atrazine D6 Diphenhydramine D3 Atrazine D5 Atrazine D5 Atrazine D5
Antihyperglycemic	Metformin Sitagliptin	Metformin D6 Sitagliptin D4
Antimalarial	Artemisinin	Atrazine D5
Antiviral/-retroviral	Nevirapine Oseltamivir	Atrazine D5 Atrazine D5
Benzodiazepine	Diazepam Oxazepam Temazepam	Diazepam D5 Oxazepam D5 Temazepam D5
Beta-blocker	Atenolol Propranolol	Atenolol D7 Propranolol D7
Calcium channel blocker	Diltiazem Verapamil	Diltiazem D3 Verapamil D7
Diuretic	Triamterene	Triamterene D5
Histamine H$_2$ receptor antagonist	Cimetidine Ranitidine	Atrazine D5 Atrazine D5
Opioid pain medication	Codeine Hydrocodone Tramadol	Codeine D6 Hydrocodone D3 Atrazine D5
Oral contraceptive	Noreistherone	Atrazine D5
Selective estrogen receptor modulator	Raloxifene	Ralixifene D4
Stimulant	Caffeine Cotinine Nicotine	Atrazine D5 Cotinine D3 Atrazine D5
β2 adrenergic receptor agonist (anti-asthma)	Salbutamol	Salbutamol D9

* Used in animal husbandry.

2.2. Sampling Kits and Water Collection Protocol

The sampling kits were designed to simplify logistics so that a large number of locations could be sampled with a minimum of effort. As the sample injection volume for the developed method was only 100 μL, the standard collection volume was set at 10 mL of sample water. Samples were collected in duplicate to provide a backup sample in case of breakage of the primary sample container during shipment. Each sampling kit therefore contained: 20 amber glass vials (15 mL) (for collection of 10 samples in duplicate), two ice packs, 10 polypropylene syringes (24 mL), 10 glass microfiber filters (0.7-μm pore size), a 500-mL stainless-steel bucket with 10-m long nylon cord attached, material to collect a field blank quality-control (QC) sample with LCMS-grade water, a standardised collection log and sample labels. Kits weighed 2.25 kg and were able to fit into a 34.5 × 21 × 14.5 cm box (approximately the size of a shoe box).

The sample collection protocol, including storage and shipping procedures, was included with the sampling kit and instructional videos were also provided online (<https://youtu.be/HeZ7xoxJXhM> and <https://youtu.be/PLvCNcVCKdc>). At each location, the bucket (included with each kit) was rinsed three times with native sample water prior to collection. Following sample collection, 20 mL of sample water was aspirated into the syringe, and the syringe filter was then attached and primed with 5 mL of sample water. Then, 5 mL of filtrate was used to rinse out a 15-mL vial prior to dispensing the remaining 10 mL of filtrate into the glass vial. This procedure was repeated once more with the same bucket of water to create the second replicate. At this point, pH, temperature or other probes may be inserted into the water remaining in the bucket to obtain additional environmental data. All vials were labelled with their location, sample date/time, and replicate number and immediately placed on ice upon collection. Prior to shipment, samples were frozen until being shipped on ice to the University of York (York, UK) Centre of Excellence in Mass Spectrometry (CoEMS) using DHL global express delivery (1–2 days). Freezing the samples prior to express shipping to CoEMS ensured the samples maintained their integrity (i.e., did not become warm during shipment) prior to analysis.

2.3. HPLC-MS/MS Protocol

The analytical method was adapted from a previously developed method for pharmaceuticals compounds [7]. Analysis occurred by direct-injection (100-µL injection volume) HPLC-MS/MS in multiple reaction monitoring (MRM) mode with positive electrospray ionisation. A Thermo Scientific Endura TSQ triple quadrupole mass spectrometer coupled with a Thermo Scientific Dionex UltiMate 3000 HPLC was used for all analyses. Two transition ions were optimised (for collision energy and retention time) in-house, one for quantitation (T1) and another for confirmation (T2) of precursor identity (Table S1). The instrument-calibrated fragmentor voltage was used for all analysis. Mobile phase A was LCMS-grade water with 0.01 M formic acid and 0.01 M ammonium formate while mobile phase B was 100% methanol. The flow rate was 0.45 mL/min. Flow was diverted away from the spectrometer for the first 1 min of the analytical run to avoid poorly retained materials (e.g., slats) from reaching the nebuliser. The HPLC gradient started at 10% B which increased to 40% at 5 min, 60% at 10 min, and 100% at 15 min, where it remained until 23 min then reduced to 10% at 23.1 min prior to a 10-min re-equilibration time between runs. Autosampler temperature was maintained at 6 °C while the column temperature was maintained at 40 °C. The collision gas was argon and was set at a pressure of 2 mTorr. Quantification occurred using a 15-point calibration prepared for 33 deuterated internal standards (Table 1, Table S1) ranging from 1 to 8000 ng/L (Table S2). All calibrants were made using a standard method as described by Furlong et al. [7] in such a way as to maintain an equal proportion of methanol in the final calibrants (Table S2). Atrazine-D5 was used where a labelled standard was not available for a specific target chemical, as established by Furlong et al. [7].

2.4. Quality Control

Extensive quality-control measures were used in-house and in the field to ensure that the laboratory and field protocols were not causing false positives or negatives in the corresponding environmental results. Materials needed to conduct one field blank were included in each sample kit which included 25 mL of LCMS-grade water, a syringe filter, syringe and two 15-mL glass vials. The procedure for collection, storage and shipment of this QC sample was exactly the same as for environmental samples, except using LCMS-grade water. This step enables an evaluation of sample contamination derived from collection in the field.

In addition to field QC measures, a blank as well as method and instrumental QCs were injected after every 10 injections during analytical runs. The laboratory blanks were pure LCMS-grade water with all internal standards spiked to a concentration of 400 ng/L. Both method and instrumental QCs consisted of all target APIs spiked in LCMS-grade water at a concentration of 80 ng/L with all QCs at 400 ng/L. However, the method QC underwent the same sample storage and preparation measures as actual samples and the instrumental QC was spiked directly into a HPLC vial prior to analysis. Before each use, the nebuliser and spray guard of the mass spectrometer were cleaned with methanol. Additionally, prior to an analytical run, the chromatography column was equilibrated with

20 injections of a composite environmental sample (made from equal aliquots of the samples being analysed in respective runs) in order to condition the chromatography column prior to analysis.

2.5. Method Validation

The method was validated based on USGS method No. O-2440-14 (National Water Quality Laboratory [NWQL] laboratory schedule 2440) for filtered water [7], which will be referred to as USGS method No. 5-B10 in this paper. Briefly, intra-/inter-day repeatability was determined at three concentrations (10, 100 and 1000 ng/L) over 3 days (*n* = 10 per concentration). Analyte response (recovery) was also determined at three concentrations (10, 100 and 1000 ng/L) in LCMS-grade water, drinking water directly from the tap (chlorinated), surface water, and WWTP influent and effluent. Surface water was obtained from the River Ouse in York City Centre (UK, GPS coordinates: 53.957397, −1.083816), drinking water was from the tap at the University of York (York, UK), and both WWTP influent and effluent were obtained from a WWTP in Barnsley, UK. Limits of detection (LOD) and quantification (LOQ) were statistically derived using the method described by Sallach et al. [15]. Briefly, respective LODs were based on the Grubbs *t*-test constant for 10 variables multiplied by the standard deviation of 10 replicate quantifications of test chemicals in mixture at the lowest calibrant level (1 ng/L). The LOQs for respective analytes were determined as two times the LOD [15]. Analytical limits were determined in LCMS-grade water, drinking water, surface water, and WWTP influent and effluent. An acceptable range for analyte response was considered between 60–130% and <20% for intra-/inter-day repeatability and precision as established by USGS method No. 5-B10 [7].

2.5.1. Evaluation of Chemical Stability

To evaluate the potential degradation of test APIs during shipment, a stability assessment was performed at three temperatures: 4 °C (*n* = 6), 20 °C (*n* = 6), and 35 °C (*n* = 6) (Table S3). Six replicates of 10-mL LCMS-grade water for each temperature were spiked with a mixture of all test chemicals to make a final mixed concentration of all the APIs at 1000 ng/L. Samples were then stored at the designated temperatures for either 2 or 7 days to provide the range of holding times from the field to the laboratory that would be encountered for this protocol. A 1-mL aliquot of each sample was collected, and analysis occurred via the same procedure as environmental samples. In addition to the stability assessment to assess if sample storage temperature during shipping affects API concentrations, the interior temperature of three sets of polystyrene packages containing two ice packs frozen at −20 °C was measured over 7 days to determine the conditions samples are likely to experience during shipment.

2.5.2. Interlaboratory Assessment

To validate the method using independent analytical results, an interlaboratory comparison was conducted where four samples (three stream samples and 1 WWTP effluent sample) were simultaneously collected for API analysis at CoEMS and the USGS using USGS method No. 5-B10 [7]. Both methods used the exact same field sample processing procedures and materials (e.g., 15-mL amber glass sample vials, 24-mL leur lock syringes, and 0.7-μm glass microfiber syringe filters) making for a more effective comparison of the analytical methods used without having added variability due to field collection and processing procedures. All samples for this interlaboratory comparison (Figure S1) were collected from Muddy Creek, North Liberty, Iowa (USA) using the provided sampling kit and designated protocols and were immediately chilled and express mailed the same day as sample collection and arriving within 24 h to the USGS NWQL in Denver (Colorado, USA) for analysis. Samples to CoEMS were frozen following collection and then express mailed where they were received within 37 h still frozen. There were 30 overlapping APIs between these two methods for this interlaboratory comparison (Table S4). USGS method No. 5-B10 uses the same chromatography column, injection volume and positive ESI as the CoEMS method, however it is conducted using an Agilent Technologies 6460 triple quadrupole tandem mass spectrometer coupled to an Agilent

1200 HPLC system [7]. The data were evaluated based on their absolute difference (%) and the order (from highest concentration to lowest) of quantified APIs.

3. Results and Discussion

3.1. Method Validation

All of the validated API parameters were assessed against the range established by USGS method No. 5-B10 (matrix recovery of 60–130% and RSD ≤20%) and the determined analytical limits (Table 2) were all in the ng/L range [7]. Mean calibration r^2-values were 0.984 ± 0.019, relative standard deviation of intra- and inter-day repeatability and precision was typically ≤20% (Table S5), and recovery from LCMS-grade water (Table S6), tap water (Table S7), surface water (Table S8), WWTP effluent (Table S9) and WWTP influent (Table S10) were typically between 60–130%. Analyte response (i.e., recovery) was comparable between LCMS-grade water, drinking water and surface water with deviation from a range of 60–130% most notable in WWTP influent (67% of determinations). This is likely due to matrix enhancement and indicates that analysis of WWTP influent may be best conducted with an initial sample clean-up method or by analysis following dilution of the sample using methods described by Furlong et al. [7]. Limits of detection were lowest in LCMS-grade water and highest in WWTP influent (Table S11). Generally, little difference was observed between the analytical limits determined in LCMS-grade water and those in tap and surface water enabling sensitive use of the method with these matrices (Table 2).

This validation indicates that the method presented here is sufficiently robust and is best used in high-throughput applications for analysis of drinking and surface water. The method can also be applied with reasonable accuracy in non-diluted WWTP effluent. While analysis of influent water is possible, dilution is recommended on a site-specific basis due to potential matrix effects. The precision of the analysis in influent samples may be improved with further sample pre-treatment such as solid phase extraction [16].

The analytical response and limits demonstrated by this method are comparable to (e.g., Furlong et al. [7], Oliveira et al. [17], Hermes et al. [18]) and more sensitive than (e.g., Campos-Mañas et al. [8]) other direct-injection HPLC-MS/MS methods. This supports similar work demonstrating that direct-injection HPLC-MS/MS can achieve robust and specific quantification at low ng/L levels without a need for sample clean-up (other than filtration) and pre-concentration [16]. Furthermore, this method provides similar environmentally relevant sensitivity as more rigorous (and likely more expensive) protocols involving sample pre-concentration and SPE (e.g., Gurke et al. [19], Paiga et al. [20]). The protocol presented here also offers a clear advantage over others as the shipment and chemical stability of the selected contaminants during shipment was also validated, as shown in Section 3.2. Therefore, use of this protocol allows for sample collection in areas geographically isolated from analytical centres, which may reduce the accessibility barriers to sensitive analytical equipment worldwide.

Table 2. Statistical overview of method validation showing the mean value and standard deviation (SD) for each parameter and the percent of determinations which fall between the acceptable range employed by U.S. Geological Survey (USGS) Method No. 5-B10.

Validation Parameter		Mean ± SD	Within Acceptable Range *
Linearity	r^2	0.984 ± 0.02	
Intra-day Repeatability (%)	10 ng/L	29 ± 19	46%
	100 ng/L	11 ± 9	82%
	1000 ng/L	8 ± 6	95%
Intermediate/Inter-day Precision (%)	10 ng/L	42 ± 41	34%
	100 ng/L	16 ± 12	72%
	1000 ng/L	9 ± 6	93%

Table 2. *Cont.*

Validation Parameter		Mean ± SD	Within Acceptable Range *
Analyte Response in LCMS-grade water (%)	10 ng/L	95 ± 32	75%
	100 ng/L	104 ± 22	90%
	1000 ng/L	94 ± 15	97%
Analyte Response in tap water (%)	10 ng/L	110 ± 27	59%
	100 ng/L	101 ± 31	80%
	1000 ng/L	106 ± 28	90%
Analyte Response in surface water (%)	10 ng/L	101 ± 88	56%
	100 ng/L	106 ± 26	82%
	1000 ng/L	106 ± 18	92%
Analyte Response in WWTP effluent (%)	10 ng/L	195 ± 212	39%
	100 ng/L	117 ± 38	59%
	1000 ng/L	108 ± 18	92%
Analyte Response WWTP influent (%)	10 ng/L	465 ± 992	20%
	100 ng/L	168 ± 81	38%
	1000 ng/L	145 ± 48	54%
Limit of Detection (ng/L)	LCMS-grade water (ng/L)	9.16	
	Drinking water (ng/L)	9.72	
	Surface water (ng/L)	11.79	
	WWTP effluent (ng/L)	20.22	
	WWTP influent (ng/L)	54.45	
Limit of Quantification (ng/L)	LCMS-grade water (ng/L)	18.32	
	Drinking water (ng/L)	19.44	
	Surface water (ng/L)	23.57	
	WWTP effluent (ng/L)	40.43	
	WWTP influent (ng/L)	108.89	

* The acceptable range for analyte response was considered between 60–130% and <20% for intra-/inter-day repeatability and precision as established by USGS method No. 5-B10 [ ].

3.2. Evaluation of Chemical Stability during Shipment

Simulated shipping events (*n* = 3) showed that the interior temperature of the polystyrene package remained below ambient temperature for 2.44 days (Figure 1). A negative relationship between chemical stability in water and both temperature and time was observed, and degradation was consistently higher with increasing temperature and time. Over 2 days, the degradation study determined a mean stability of 92% ± 8.6% at 4 °C, 89% ± 7.9% at 20 °C and 90% ± 13% at 35 °C. After 7 days at respective temperatures, stability dropped to 83% ± 11%, 80% ± 13% and 77% ± 19% at 4 °C, 20 °C and 35 °C respectively (Figure 2, Table S3).

Fluoxetine and its metabolite norfluoxetine were the least stable, exhibiting a mean degradation of 29.7–37.5% and 29.3–34.2% respectively over 2 days and 41.2–50.3% and 41.6–49.2% respectively over 7 days (Table S3). Interestingly, previous work has indicated that both fluoxetine and norfluoxetine are relatively hydrolytically stable [21]. However, they are known to rapidly partition to solid matrices such as sediment [22]. Hence, the decreased stability of these chemicals observed here may partially be an artefact of sorption to test materials including the glass test vessel/PTFE cap or plastic pipette tips. The tetracycline antibiotics tetracycline and oxytetracycline were also relatively unstable exhibiting mean degradation of 20.9–24.2% and 21.1–29.3% respectively after 2 days and 34.8–43.1% and 32.3–39.3% respectively after 7 days (Table S3). Previous work has indicated that both tetracycline and oxytetracycline are quickly degradable via oxidation (e.g., Jeong et al. [23]). Despite the observed degradation, the dissipation of fluoxetine, norfluoxetine and the tetracyclines were still of an acceptable level to draw conclusions over the relative fate and abundance of these contaminants in water. All remaining test chemicals generally demonstrated degradation of <20% over 2 days and <30% over 7 days (Table S3).

As express shipping usually takes 1–2 days and samples are shipped frozen, it is unlikely that any API in this method would significantly degrade during shipment. Here, a mean degradation rate of

11% ± 7.9% (median degradation of 9.1%) over 2 days at 20 °C (Table S3) is superior to the typical loss of 20–40% on sample pre-treatment steps not required by this method (e.g., solid phase extraction) from water at environmentally relevant pH levels [20,24].

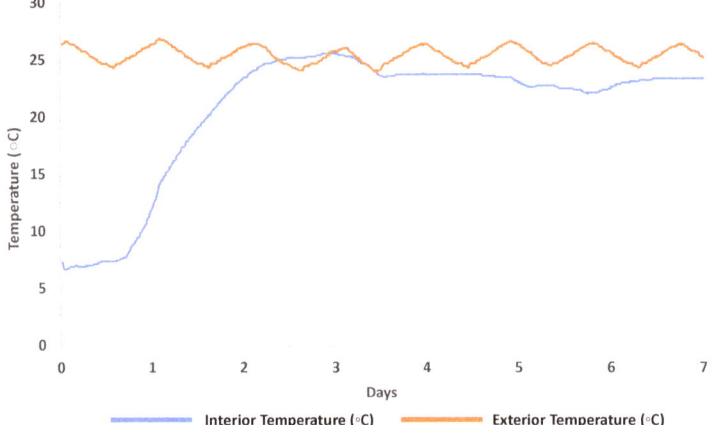

Figure 1. Mean (*n* = 3) interior and exterior temperatures of a sample shipping package containing two ice packs (as standardised for all sampling kits) over 7 days.

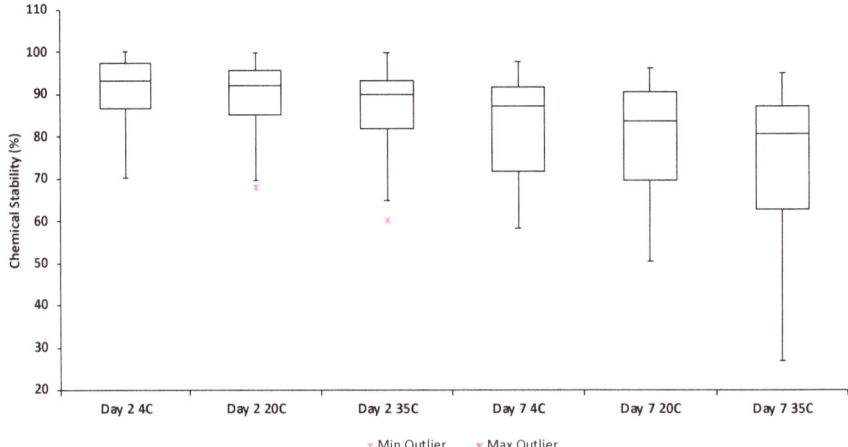

Figure 2. Stability of the target chemicals in liquid chromatography-mass spectrometry (LCMS)-grade water held at 4 °C, 20 °C and 35 °C for 2 and 7 days respectively.

3.3. Interlaboratory Assessment

The 30 overlapping APIs between the CoEMS and USGS methods led to 120 chemical determinations for comparison, with chemical results being confirmed in 98% of determinations (117 of 120, Table S4). Only three determinations for codeine were confirmed at low concentrations (i.e., 15 to 26 ng/L) via the USGS method while not detected by the CoEMS method (Table S4). Of the 120 total determinations, 48 were nondetects in both methods, 14 were not detected in one method but confirmed between the LOD and LOQ of the other (e.g., mainly due to differences between LODs), 55 determinations were confirmed and quantified in both methods, while three were confirmed and quantified in one method but not in the other (Table S4).

The 55 determinations which were confirmed and quantified by both methods covered 19 APIs (Figure 3). Concentrations determined between the methods were in agreement with an overall mean deviation of 19.5% ± 12.3% (Figure 3). Within this deviation, 76.4% of the determinations made at CoEMS were lower than those determined by the USGS (mean 19.6% ± 9.6% lower). No substantial difference was observed between the interlaboratory deviations in WWTP effluent and surface water with median discrepancies of 19.7% (31.5 ng/L) and 19.4% (37.9 ng/L) respectively. The absolute differences in respective matrices ranged from 2.4% (1 ng/L, trimethoprim) to 38.7% (31.5 ng/L, sitagliptin) in effluent and 0.8% (2.6 ng/L, tramadol) to 36.4% (104 ng/L, lidocaine) in surface water (Figure 3). Generally, deviation between the two values (Table S4) was within the intra-day repeatability of the CoEMS method (Table S5).

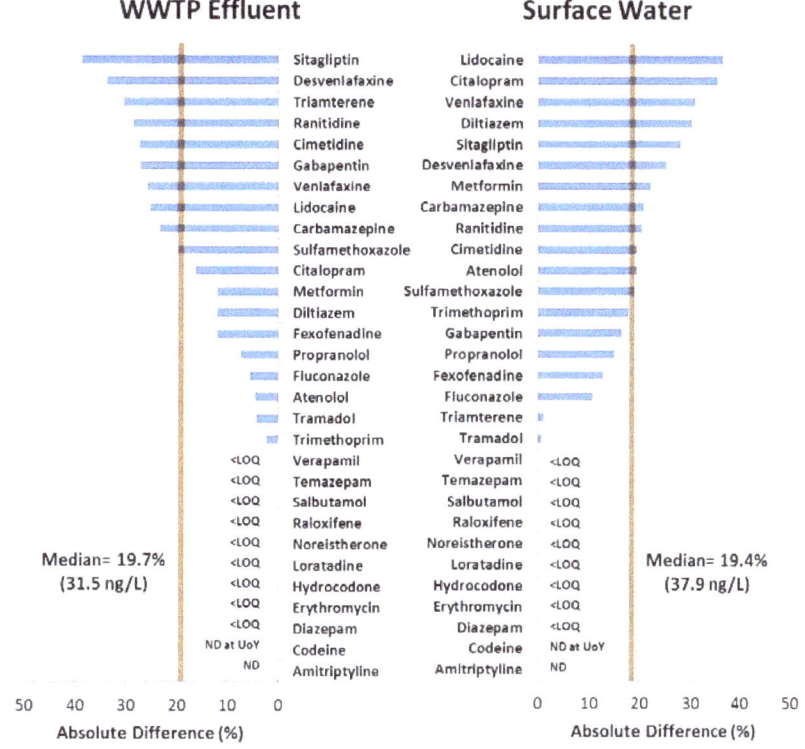

Figure 3. Relative percent difference (RPD) between concentrations of 30 medicinal chemicals (present in both respective methods) quantified by the USGS (using USGS method No. 5-B10 [7]) and CoEMS (using the method presented in this work) in both WWTP effluent and surface water with median RPD represented by the vertical bar.

The top five APIs prioritised by concentration in both WWTP effluent and surface water were identical for both the USGS and CoEMS assessments: Fexofenadine, gabapentin, metformin, desvenlafaxine and venlafaxine (Figure 4).

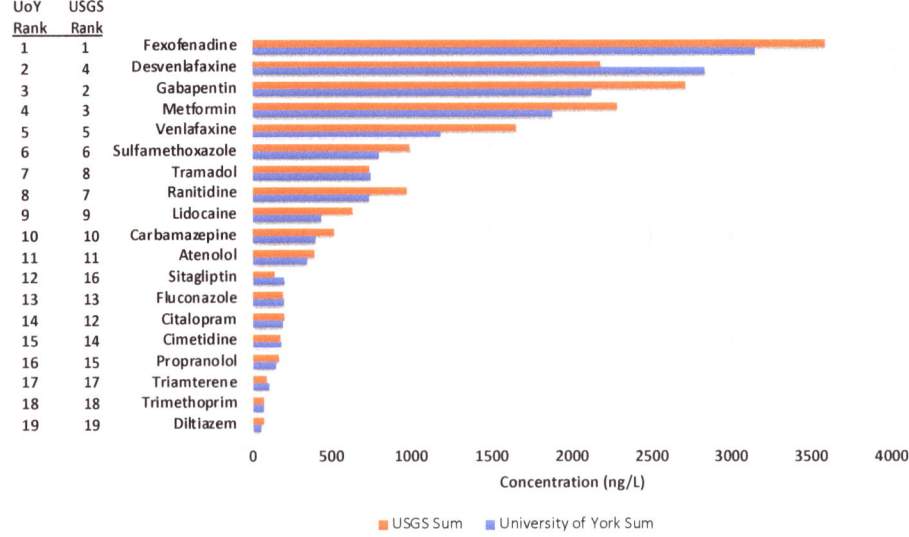

Figure 4. Sum concentrations of the 19 quantified active pharmaceutical ingredients (APIs) determined by analysis at the CoEMS and USGS respectively with their prioritisation rank determined by total concentration.

3.4. Concentrations of Studied APIs in Muddy Creek

Of the 61 APIs in the presented method, 31 were detected in at least one sample from Iowa (Table S12). The highest number of APIs was detected in the WWTP effluent sample (*n* = 31) and the lowest (*n* = 16) in the stream sample collected upstream of the WWTP discharge point. The fluoroquinolones ciprofloxacin and enrofloxacin were the only APIs detected above quantification levels in the upstream sample (54.3 and 185.6 ng/L respectively), potentially due to upstream agricultural pressures or other urban sources (e.g., leaking sewer lines or septic systems). The total API concentration was lowest upstream of the WWTP (239.9 ng/L), highest in the WWTP effluent (10,373 ng/L) and attenuated (from 5561.5 to 3699.4 ng/L) with increasing distance downstream from the WWTP outfall (Figure 5) as one would expect in an effluent-dominated system [25]. At the time of sampling, 55% of Muddy Creek flow downstream from the WWTP consisted of sewage plant effluent. The composition of the detected APIs in Muddy Creek was clearly defined by those in the WWTP effluent and was dominated by antihistamines > antidepressants > antiepileptics > and antihyperglycemics (Figure 5). The API detected at the highest concentration was the antihistamine fexofenadine at a maximum concentration of 1644 ng/L in the effluent and 926 ng/L in surface water collected downstream of the WWTP (Table S4). This was closely followed by the antidepressant desvenlafaxine in the WWTP effluent with a concentration of 1617 ng/L. Interestingly, this chemical is also an active metabolite of the antidepressant venlafaxine, which was detected at 681.3 ng/L in the same effluent sample.

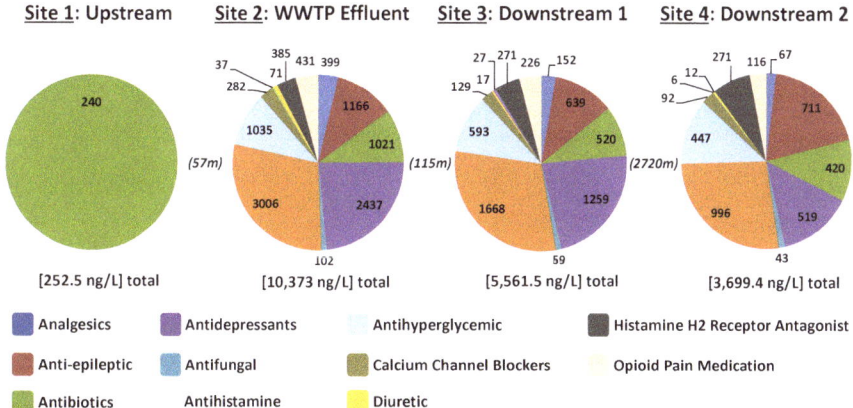

Figure 5. Muddy Creek study sites by proportion of detected therapeutic class of the 61 target APIs in the presented method with [site total concentration] and (*distance between sites in meters*). Pie charts are labelled with the total detected concentration (ng/L) of APIs belonging to respective therapeutic classes.

3.5. Implication for Future Research

The method presented here offers the first truly transferable sampling, shipping and mass spectrometry protocol for use in global monitoring campaigns. Potential applications include research extending the spatial and temporal resolution of published data on API contamination of water globally, applications related to antimicrobial resistance selection, as well as the potential for further method development enabling determination of other aquatic contaminants including persistent organic pollutants and metals. The small size of the sample vials and corresponding shipping packages and ease of sample collection may encourage the incorporation of citizen science in future API research. Fast, easy and cheap analytical methods are key to advancing the understanding of exposure to chemical contaminants via water and this method offers a clear step forward.

Supplementary Materials: Supplementary tables and figures are available online at: http://www.mdpi.com/2076-3417/9/7/1368/s1. Figure S1: Map of the sampling locations along Muddy Creek with GPS coordinates and sample site description, Table S1: Mass spectrometer conditions and target ions with precursor, transition 1 (T1) for quantitation and transition 2 (T2) for confirmation mass-to-charge ratios, collision energy and retention times, Table S2: Concentrations of the 15 calibration levels used in this method with the volumes and concentrations of mixed stock solutions used to create final calibrants with an equal amount of organic solvent in each, Table S3: Stability of monitored chemicals over 2 and 7 days at three temperatures, Table S4: Concentrations of 30 medicinal chemicals quantified in Muddy Creek (Iowa, USA) by the USGS (Central Midwest Water Science Center, Iowa City, Iowa) and University of York (York, UK), Table S5: Linearity (n = 3) and relative standard deviation (%) of both Intra-day repeatability (n = 6) and intermediate precision (n = 6) for the 61 APIs in the presented method validation with cells shaded in red indicating RSD > 20%, Table S6: Recovery (%) of 61 APIs from LCMS-grade water (n = 10) with cells shaded in green indicating a recovery between 60–130%, Table S7: Recovery (%) of 61 APIs from drinking (tap) water obtained from the University of York (n = 10) with cells shaded in green indicating a recovery between 60–130%, Table S8: Recovery (%) of 61 APIs from surface water obtained from the River Ouse in York City Centre, UK (n = 10) with cells shaded in green indicating a recovery between 60–130%, Table S9: Recovery (%) of 61 APIs from wastewater treatment plant (WWTP) effluent obtained from a WWTP in Barnsley, UK (n = 10) with cells shaded in green indicating a recovery between 60–130%, Table S10: Recovery (%) of 61 APIs from wastewater treatment plant (WWTP) influent obtained from a WWTP in Barnsley, UK (n = 10) with cells shaded in green indicating a recovery between 60–130%, Table S11: Limits of Detection (LOD) and Quantification (LOQ) determined in respective matrices, Table S12: Concentrations, in ng/L, of all 61 APIs determined by the method presented in this paper found in the Muddy Creek (North Liberty, Iowa, USA).

Author Contributions: J.L.W.: Co-conception of the idea, sample analysis, data analysis, data interpretation, writing of the manuscript and editing; A.B.A.B.: Co-conception of the idea, data interpretation, writing of the manuscript and editing. D.W.K.: Sample collection/analysis, data analysis, data interpretation and editing.

Funding: This research received no external funding.

Acknowledgments: The authors would like to thank Shannon Meppelink and Jessica Garret (USGS Central Midwest Water Science Center) for the collection of stream and WWTP effluent samples at the Muddy Creek sites and programmatic support by the Toxic Substances Hydrology Program of the USGS Environmental Health Mission Area. Any use of trade, product, or firm names is for descriptive purposes only and does not imply endorsement by the U.S. Government. The authors also acknowledge the York CoEMS which was created thanks to a major capital investment through Science City York, supported by Yorkshire Forward with funds from the Northern Way Initiative, and subsequent support from EPSRC (EP/K039660/1; EP/M028127/1).

Conflicts of Interest: The authors declare no conflict of interest.

References

1. aus der Beek, T.; Weber, F.A.; Bergmann, A.; Hickmann, S.; Ebert, I.; Hein, A.; Küster, A. Pharmaceuticals in the environment—Global occurrences and perspectives. *Environ. Toxicol. Chem.* **2016**, *35*, 823–835. [CrossRef]
2. Parthasarathy, R.; Monette, C.E.; Bracero, S.; Saha, M.S. Methods for field measurement of antibiotic concentrations: Limitations and outlook. *FEMS Microbiol. Ecol.* **2018**, *94*, 105. [CrossRef] [PubMed]
3. Tisler, S.; Zwiener, C. Formation and occurrence of transformation products of metformin in wastewater and surface water. *Sci. Total Environ.* **2018**, *628*, 1121–1129. [CrossRef]
4. Hughes, S.R.; Kay, P.; Brown, L.E. Global synthesis and critical evaluation of pharmaceutical data sets collected from river systems. *Environ. Sci. Technol.* **2012**, *47*, 661–677. [CrossRef] [PubMed]
5. Van den Brink, P.J.; Boxall, A.B.; Maltby, L.; Brooks, B.W.; Rudd, M.A.; Backhaus, T.; Spurgeon, D.; Verougstraete, V.; Ajao, C.; Ankley, G.T.; et al. Toward sustainable environmental quality: Priority research questions for Europe. *Environ. Toxicol. Chem.* **2018**, *37*, 2281–2295. [CrossRef] [PubMed]
6. Hossain, A.; Nakamichi, S.; Habibullah-Al-Mamun, M.; Tani, K.; Masunaga, S.; Matsuda, H. Occurrence and ecological risk of pharmaceuticals in river surface water of Bangladesh. *Environ. Res.* **2018**, *165*, 258–266. [CrossRef]
7. Furlong, E.T.; Noriega, M.C.; Kanagy, C.J.; Kanagy, L.K.; Coffey, L.J.; Burkhardt, M.R. Determination of Human-Use Pharmaceuticals in Filtered Water by Direct Aqueous Injection: High-Performance Liquid Chromatography/Tandem Mass Spectrometry. U.S. Geological Survey Techniques and Methods 5-B10. 2014. Available online: https://pubs.usgs.gov/tm/5b10/pdf/tm10-b5.pdf (accessed on 8 March 2019).
8. Campos-Mañas, M.C.; Plaza-Bolaños, P.; Sánchez-Pérez, J.A.; Malato, S.; Agüera, A. Fast determination of pesticides and other contaminants of emerging concern in treated wastewater using direct injection coupled to highly sensitive ultra-high performance liquid chromatography-tandem mass spectrometry. *J. Chromatogr. A* **2017**, *1507*, 84–94. [CrossRef]
9. Chiaia, A.C.; Banta-Green, C.; Field, J. Eliminating solid phase extraction with large-volume injection LC/MS/MS: Analysis of illicit and legal drugs and human urine indicators in US wastewaters. *Environ. Sci. Technol.* **2008**, *42*, 8841–8848. [CrossRef]
10. Bayen, S.; Yi, X.; Segovia, E.; Zhou, Z.; Kelly, B.C. Analysis of selected antibiotics in surface freshwater and seawater using direct injection in liquid chromatography electrospray ionization tandem mass spectrometry. *J. Chromatogr. A* **2014**, *1338*, 38–43. [CrossRef]
11. Boxall, A.B.; Rudd, M.A.; Brooks, B.W.; Caldwell, D.J.; Choi, K.; Hickmann, S.; Innes, E.; Ostapyk, K.; Staveley, J.P.; Verslycke, T.; et al. Pharmaceuticals and personal care products in the environment: What are they key questions? *Environ. Health Perspect.* **2012**, *120*, 1221–1229. [CrossRef]
12. Wellington, E.M.; Boxall, A.B.; Cross, P.; Feil, E.J.; Gaze, W.H.; Hawkey, P.M.; Johnson-Rollings, A.S.; Jones, D.L.; Lee, N.M.; Otten, W.; et al. The role of the natural environment in the emergence of antibiotic resistance in Gram-negative bacteria. *Lancet Infect. Dis.* **2013**, *13*, 155–165. [CrossRef]
13. Roca, I.; Akova, M.; Baquero, F.; Carlet, J.; Cavaleri, M.; Coenen, S.; Cohen, J.; Findlay, D.; Gyssens, I.; Heure, O.E.; et al. The global threat of antimicrobial resistance: Science for intervention. *New Microbes New Infect.* **2015**, *6*, 22–29. [CrossRef] [PubMed]
14. Brodin, T.; Piovano, S.; Fick, J.; Klaminder, J.; Heynen, M.; Jonsson, M. Ecological effects of pharmaceuticals in aquatic systems—Impacts through behavioural alterations. *Philos. Trans. R. Soc. B Biol. Sci.* **2014**, *369*, 20130580. [CrossRef]
15. Sallach, J.B.; Snow, D.; Hodges, L.; Li, X.; Bartelt-Hunt, S. Development and comparison of four methods for the extraction of antibiotics from a vegetative matrix. *Environ. Toxicol. Chem.* **2016**, *35*, 889–897. [CrossRef]

16. Bisceglia, K.J.; Jim, T.Y.; Coelhan, M.; Bouwer, E.J.; Roberts, A.L. Trace determination of pharmaceuticals and other wastewater-derived micropollutants by solid phase extraction and gas chromatography/mass spectrometry. *J. Chromatogr. A* **2010**, *1217*, 558–564. [CrossRef] [PubMed]

17. Oliveira, T.S.; Murphy, M.; Mendola, N.; Wong, V.; Carlson, D.; Waring, L. Characterization of pharmaceuticals and personal care products in hospital effluent and waste water influent/effluent by direct-injection LC-MS-MS. *Sci. Total Environ.* **2015**, *518*, 459–478. [CrossRef] [PubMed]

18. Hermes, N.; Jewell, K.S.; Wick, A.; Ternes, T.A. Quantification of more than 150 micropollutants including transformation products in aqueous samples by liquid chromatography-tandem mass spectrometry using scheduled multiple reaction monitoring. *J. Chromatogr. A* **2018**, *1531*, 64–73. [CrossRef] [PubMed]

19. Gurke, R.; Rossmann, J.; Schubert, S.; Sandmann, T.; Rößler, M.; Oertel, R.; Fauler, J. Development of a SPE-HPLC–MS/MS method for the determination of most prescribed pharmaceuticals and related metabolites in urban sewage samples. *J. Chromatogr. B* **2015**, *990*, 23–30. [CrossRef] [PubMed]

20. Paiga, P.; Santos, L.H.M.L.M.; Delerue-Matos, C. Development of a multi-residue method for the determination of human and veterinary pharmaceuticals and some of their metabolites in aqueous environmental matrices by SPE-UHPLC–MS/MS. *J. Pharm. Biomed. Anal.* **2017**, *135*, 75–86. [CrossRef] [PubMed]

21. Yin, L.; Ma, R.; Wang, B.; Yuan, H.; Yu, G. The degradation and persistence of five pharmaceuticals in an artificial climate incubator during a one year period. *RSC Adv.* **2017**, *7*, 8280–8287. [CrossRef]

22. Kwon, J.W.; Armbrust, K.L. Laboratory persistence and fate of fluoxetine in aquatic environments. *Environ. Toxicol. Chem.* **2006**, *25*, 2561–2568. [CrossRef] [PubMed]

23. Jeong, J.; Song, W.; Cooper, W.J.; Jung, J.; Greaves, J. Degradation of tetracycline antibiotics: Mechanisms and kinetic studies for advanced oxidation/reduction processes. *Chemosphere* **2010**, *78*, 533–540. [CrossRef] [PubMed]

24. Leusch, F.D.L.; Prochazka, E.; Tan, B.L.L.; Carswell, S.; Neale, P.; Escher, B.I. Optimising micropollutants extraction for analysis of water samples: Comparison of different solid phase materials and liquid-liquid extraction. In Proceedings of the Science Forum and Stakeholder Engagement: Building Linkages, Collaboration and Science quality, Brisbane, QL, Australia, 19–20 June 2012; pp. 191–195.

25. Bradley, P.M.; Barber, L.B.; Duris, J.W.; Foreman, W.T.; Furlong, E.T.; Hubbard, L.E.; Hutchinson, K.J.; Keefe, S.H.; Kolpin, D.W. Riverbank filtration potential of pharmaceuticals in a wastewater-impacted stream. *Environ. Pollut.* **2014**, *193*, 173–180. [CrossRef] [PubMed]

Article

Quantitative Occurrence of Antibiotic Resistance Genes among Bacterial Populations from Wastewater Treatment Plants Using Activated Sludge

Adriana Osińśka [ID], Ewa Korzeniewska *[ID], Monika Harnisz [ID] and Sebastian Niestępski

Department of Environmental Microbiology, Faculty of Environmental Sciences,
University of Warmia and Mazury in Olsztyn, Prawocheńskiego 1, 10-720 Olsztyn, Poland;
adriana.osinska@uwm.edu.pl (A.O.); monika.harnisz@uwm.edu.pl (M.H.);
sebastian.niestepski@uwm.edu.pl (S.N.)
* Correspondence: ewa.korzeniewska@uwm.edu.pl; Tel.: +48-89-523-37-52

Received: 28 December 2018; Accepted: 20 January 2019; Published: 23 January 2019

Abstract: Wastewater treatment plants (WWTPs) are an important reservoir in the development of drug resistance phenomenon and they provide a potential route of antibiotic resistance gene (ARGs) dissemination in the environment. The aim of this study was to assess the role of WWTPs in the spread of ARGs. Untreated and treated wastewater samples that were collected from thirteen Polish WWTPs (applying four different modifications of activated sludge–based treatment technology) were analyzed. The quantitative occurrence of genes responsible for the resistance to beta-lactams and tetracyclines was determined using the real-time PCR method. Such genes in the DNA of both the total bacterial population and of the *E. coli* population were analyzed. Among the tested genes that are responsible for the resistance to beta-lactams and tetracyclines, bla_{OXA} and bla_{TEM} and tetA were dominant, respectively. This study found an insufficient reduction in the quantity of the genes that are responsible for antibiotic resistance in wastewater treatment processes. The results emphasize the need to monitor the presence of genes determining antibiotic resistance in the wastewater that is discharged from treatment plants, as they can help to identify the hazard that treated wastewater poses to public health.

Keywords: wastewater; qPCR; tetracyclines; beta-lactams; ARGs; *Escherichia coli*

---

1. Introduction

The occurrence and spread of antibiotic-resistant bacteria (ARB) and antibiotic resistance genes (ARGs) is a serious health protection problem worldwide [1]. According to the World Health Organization, the occurrence of antibiotic resistance among bacterial populations is regarded as one of the major hazards and challenges to public health in the 21st century [2], since it has serious economic consequences as well as risks to the health and lives of both humans and animals [3]. A major problem in controlling this phenomenon is the lack of national or international legal regulations controlling the spread of ARB and ARGs in the environment [4]. The existing reports and definitions of drug resistant bacteria refer only to clinical strains, while similar characteristics for environmental strains are still missing [5]. The control of the aquatic environment appears particularly important, as it provides the main spread route of ARB and ARGs [6]. Even though the European Union (EU) Water Framework Directive contains a provision on ensuring the good quality of waters according to specific standards, it does not refer to the phenomenon of antibiotic resistance. The genes that are responsible for resistance to antibiotics are regarded as environmental pollutants [7] that are capable of spreading among bacteria in natural environments and drinking water resources [8,9]. This is due to the fact that non-pathogenic bacterial species with antibiotic resistance genes may serve as a source of antibiotic resistance genes for pathogenic bacteria [10]. ARGs that are found among bacterial pathogens largely originate from the environment, which receives treated or untreated wastewater generated by animal farms, aquaculture, and industry [9,11].

One of the main reservoirs of ARB and ARG pollutants and the source of their spread in the natural environment [12–14] are wastewater treatment plants (WWTPs) and treated wastewater discharged to surface water bodies [5,6,15–17]. WWTPs receiving high concentrations of microbial contaminants with wastewater from hospitals, agriculture, and industry stimulate the transfer of genetic information between pathogenic and environmental microorganisms. In addition, the conditions prevailing in wastewater treatment plants, such as a high content of microorganism populations, the relative abundance of nutrients, and the presence of sub-threshold levels of antibiotic substances in wastewater [16], provide an environment favourable for the survival of ARB and the transfer of ARGs.

The increase in microbial diversity and the number of mobile genetic elements [18] and ARGs [19] is significantly affected by the activated sludge. Microorganisms forming particles of sludge are responsible for the accumulation and release of bacterial plasmids carrying ARGs [20], which facilitate the exchange of these genes between bacteria. Another factor contributing to the development of antibiotic resistance phenomenon during the wastewater treatment process is the presence of antibiotics, which, even at low concentrations, may induce genetic responses leading to adaptations and mutations among the microbial population [21]. Therefore, conventional activated sludge-based wastewater treatment methods provide ideal conditions for the transfer of these genes [22], which was confirmed by many authors [23–25]. The results of these studies show that the bacteria released with the discharged treated wastewater may be capable of active transmission of resistance genes among environmental microorganisms. Bengtsson-Palme et al. [26] demonstrated that the number of bacteria decreased following the wastewater treatment processes, while the abundance and diversity of genes determining antibiotic resistance did not change significantly. Moreover, a comparison of treated and untreated wastewater samples demonstrated that certain resistance genes, e.g., carbapenemase OXA-48 and the count of mobile genetic elements, were also not significantly reduced [6,26,27].

WWTPs are not specifically designed to remove the antibiotics ARB and ARGs [13]. The main requirement imposed on WWTPs is to ensure the optimum values of organic matter, nitrogen, and phosphorus, since the discharge of wastewater containing high levels of these elements may contribute to oxygen depletion and an increase in the trophic state of the receiving waters. However, the penetration of microbial contaminants with the treated wastewater is not usually subject to regulations or monitoring. It should be mentioned that, despite the reduction in the total number of bacteria and ARB in the wastewater treatment process, large numbers of bacteria exhibiting multi-drug-resistance characterized by higher virulence could still penetrate into the environment with the wastewater [6,28].

Since the presence of antibiotic resistant bacteria and antibiotic resistance genes in the environment poses a potential hazard to health, the present study applied molecular analyses for an assessment of the role of wastewater treatment plants in the spread of ARGs. Quantitative testing of genes encoding the resistance to beta-lactam antibiotics and tetracyclines (antibiotic classes most commonly used in medical treatment) enabled a performance assessment of a range of wastewater treatment plants that apply various modifications of the activated sludge-based treatment process. Genomic DNA samples originating from the population cultured in the study were used. A significant limitation of this method is the small amount of growth-capable environmental bacteria possible to obtain, although the aim of the study was to analyse antibiotic resistance in live bacterial cells with ARGs. The presence of live bacterial cells that are capable of multiplying constitutes a considerably greater potential threat of transmission of these genes from WWTPs to the environment. The study results may be of significance in identifying the degree of risks to the public health of humans and animals exploiting water bodies that receive treated wastewater.

2. Materials and Methods

2.1. Study Sites and Sampling

For the purposes of the study, untreated and treated wastewater samples were collected from thirteen wastewater treatment plants of various capacity, hydraulic retention time (HRT), and characteristics of the untreated wastewater, located in the Warmińsko-Mazurskie District in Poland. All of the samples were collected after the secondary treatment process. No additional wastewater disinfection was performed in the analysed WWTPs. Wastewater treatment plants were divided into four categories according to the applied modification of the wastewater treatment system: A—WWTPs (No. I, II) with an anaerobic-anoxic-aerobic (A_2O) bio-reactor system; B—WWTPs (No. III, IV, V, VI, VII) with a mechanical and biological system; C—WWTPs (No. VIII, IX, X) with SBRs (Sequencing Batch Reactors); and, D—WWTPs (No. XI, XII, XIII) with an increased nutrient removal mechanical and biological system (see Supplementary Material, Table S1). Wastewater samples were collected in the winter (February) to sterile bottles, transported to the laboratory at a temperature of 4 °C, and subjected to analysis on the day of collection [29]. The sample collection period was selected to correspond with an increased morbidity incidence and, thus, with higher antibiotic consumption.

2.2. Physicochemical Parameters and Number of Antibiotic-Resistant Bacteria

Physicochemical parameters of the wastewater samples under study, including five-day Biological Oxygen Demand (BOD), Chemical Oxygen Demand (COD), and Total Suspended Solids (TSS), were assessed simultaneously with a microbiological analysis. The total number of bacteria as well as bacteria resistant to β-lactams (amoxicillin, cefotaxime) and tetracyclines (oxytetracycline, doxycycline) and the total number of *Escherichia coli* and *Escherichia coli* resistant to the same drugs were determined according to procedures that were described by Osińska et al. [29].

2.3. Bacterial Inoculation and DNA Extraction

To obtain 20–80 colony forming units (CFU) per plate, untreated (UWW) and treated wastewater (TWW) samples were decimally diluted with saline water and passed through a cellulose filter (pore diameter of 0.45 μM, Millipore, Merck KGaA, Darmstadt, Germany). Greater accuracy was achieved by plating in triplicates. The total number of bacteria and bacteria resistant to β-lactams (amoxicillin, cefotaxime) and tetracyclines (oxytetracycline, doxycycline), as well as the total number of *Escherichia coli* and *Escherichia coli* resistant to the same drugs were determined on plates with TSA medium (Oxoid, Thermo Fisher Scientific, Carlsbad, USA) and the mFc Agar medium (Merck, Merck KGaA, Darmstadt, Germany) with/without antibiotic supplementation, respectively. The plates were incubated at 30 °C for 48 h. *E. coli* were cultured at 44.5 ± 0.2 °C for 24 h. The filters were then transferred to sterile screw cap tubes (50 mL), and 30 mL of 1 × PBS was added to the tubes. They were shaken (200 rpm/min, three hours) at room temperature. The entire precipitate was then transferred to 2.0 mL Eppendorf tubes and centrifuged (9000 rpm/min, 15 min). DNA extraction from a bacterial pellet was then performed using isolation kits (Genomic Mini A&A Biotechnology kit, Gdynia, Poland), according to the manufacturer's instructions. The concentration and quality of the extracted DNA was determined by microspectrophotometry (NanoDrop® ND-1000, Nano Drop Technologies, Willmington, DE, USA). The DNA was stored at −20 °C for further analysis.

2.4. Quantification Analysis of ARGs

The obtained DNA samples originating from antibiotic resistant bacteria were used for ARG analysis with a conventional polymerase chain reaction (PCR) and a real-time polymerase chain reaction (qPCR) using specific primers. During preliminary testing, the occurrence of seven genes responsible for the resistance to tetracyclines (*tet*A, *tet*C, *tet*L, *tet*M, *tet*O, *tet*A(P), *tet*X) and ten genes responsible for the resistance to beta-lactams (*bla*$_{TEM}$, *bla*$_{SHV}$, *bla*$_{OXA}$, *bla*$_{CTX-M}$, *bla*$_{CTX-M-1}$, *bla*$_{CTX-2}$, *bla*$_{CTX-M-9}$, *bla*$_{VEB}$, *bla*$_{CMY}$, *bla*$_{AMP-C}$) were investigated using the PCR reaction. These genes were selected based on previous studies of the authors [11,28,30]. For ARG quantitative analysis with qPCR, genes (bla_{TEM}, bla_{SHV}, bla_{OXA}, $tetA$, $tetM$) were selected based on their occurrence in environmental samples in the preliminary testing. Moreover, the presence of gene $uidA$ in *E. coli* was also used as a marker of the taxonomic genotype. Antibiotic resistance genes were used to create standard curves derived from wastewater strains. The standard curves were generated by cloning the amplicon from the positive control into vector pCR2.1-TOPO (Invitrogen, Massachusetts, USA). All qPCR reactions were carried out using a Roche Light Cycler 480 device (Roche Applied Science, Denver, CO, USA). The reactions were carried out in triplicate to ensure repeatability, using both the negative and positive control.

2.4.1. Preparation of Standard Curves

Prior to the preparation of serial dilutions to create the standard curve, it was necessary to calculate the number of copies of the gene under study. For this purpose, the following equation was applied:

$$\text{number of copies} = \frac{(\text{amount of DNA [ng]}) \times 6.022 \times 10^{23}}{\text{average size of genome [bp]} \times 10^9 \times 650} \tag{1}$$

where: 650 g/mole means average molecular weight per base pair, 6.222×10^{23}/mole is Avogadro's number, including molecules, and 10^9 ng/g is the conversion factor [29]. Standard curves for qPCR were obtained within the range from 10^8 to 10^2 gene copies/µL. The negative control sample was distilled-deionised water (ddH$_2$O) added to the qPCR reaction mixture as a substitute for matrix DNA.

2.4.2. qPCR Reaction Conditions

All of the analysed genes were quantitatively determined with the qPCR method using SYBR Green (Roche Applied Science, Denver, CO, USA). Each sample was analysed in triplicate. After each qPCR assay, a melt curve was constructed by increasing the temperature from 65 to 95 °C to verify the specificity. The point of intersection of the fluorescence signal with the threshold line determined the cycle threshold (Ct).

All of the primers had been previously validated (for primer sequences, amplicon sizes, annealing temperatures, references for each sequence, and additional details regarding qPCR conditions see Supplementary Material, Table S2.

2.5. Data Analyses

Statistical analyses were carried out using the STATISTICA 10 software package (StatSoft Inc., 1984–2011). The *p* value of <0.05 was adopted as indicating significance. The microorganism count, gene concentration, and Spearman's rank correlation were analysed to identify correlations between the physicochemical parameters and HRT values for the treatment plants. Moreover, an ANOVA test was performed in order to identify the differences between the gene concentration in the DNA of populations originating from both treated and untreated wastewater from the analysed WWTPs.

3. Results

3.1. Concentration of Physicochemical Parameters and Number of Antibiotic-Resistant Bacteria

The physicochemical parameter values of the analysed wastewater samples and the total number of antibiotic-resistant bacteria and antibiotic resistant *E. coli* were published by Osińska et al. [29]. The average BOD values in the untreated wastewater samples ranged from 401 to 606 mg/L and following the wastewater treatment process they were reduced by 97–99.8%. Meanwhile, the observed COD values ranged from 1.199 to 1.540 mg/L in the UWW samples and, following the treatment process, their levels were reduced by 92–98%. The average total number of antibiotic-resistant bacteria in UWW ranged from 7.04×10^5 to 1.6×10^7 CFU/mL and from 3.32×10^4 to 2.75×10^6 CFU/mL for bacteria resistant to beta-lactams and bacteria resistant to tetracyclines, respectively. These values were reduced by 68–99.9%, although the average count of antibiotic-resistant *E. coli* in untreated wastewater ranged from 3.67×10^2 to 4.45×10^5 and, following the treatment, it was reduced by 0–99.9%. Despite the high reduction in the antibiotic-resistant bacteria count, the percentage of these bacteria in the total number of microorganisms increased following the treatment process. The COD and BOD values were directly positively correlated with the number of analysed microorganisms ($p < 0.05$) (Table 1).

Table 1. Correlations between gene concentrations, total bacteria counts, WWTP parameters, and basic physico-chemical parameters.

	TSA/ 1 mL	TSA-AMO/ 1 mL	TSA-CTX/ 1 mL	TSA-OX/ 1 mL	TSA-DOX/ 1 mL	*tet*A	*tet*M	*bla*TEM	*bla*OXA	*bla*SHV	*uid*A	HRT	TSS	BOD
								Spearman's Rank Coefficient						
TSA-AMO/1 mL	0.95													
TSA-CTX/1 mL	0.94	0.94												
TSA-OX/1 mL	0.90	0.92	0.89											
TSA-DOX/1 mL	0.85	0.92	0.90	0.87										
*tet*A	0.22	0.30	0.28	0.26	0.38									
*tet*M	0.20	0.21	0.28	0.20	0.29	0.13								
*bla*TEM	**0.42**	0.34	0.36	0.27	0.30	0.47	0.03							
*bla*OXA	0.17	0.16	0.20	0.19	0.21	0.02	0.44	0.15						
*bla*SHV	0.41	0.37	0.42	0.32	0.46	0.47	0.33	0.42	0.42					
*uid*A	−0.01	0.00	0.03	−0.04	0.03	0.39	0.11	0.68	0.22	0.26				
HRT	−0.05	0.00	−0.02	−0.11	0.01	0.41	0.10	0.01	−0.28	0.31	0.39			
TSS	0.03	−0.04	−0.09	0.06	−0.04	−0.37	−0.15	0.29	0.08	0.08	**−0.40**	**−0.49**		
BOD	**−0.68**	**−0.72**	**−0.71**	**−0.62**	**−0.66**	−0.01	−0.27	−0.24	−0.18	−0.18	0.00	0.04	0.00	
COD	−0.30	−0.35	−0.34	**−0.43**	−0.38	**−0.45**	0.11	−0.18	−0.08	−0.15	0.30	0.19	−0.14	0.12

TSA—total bacteria counts, TSA+AMO—amoxicillin-resistant bacteria counts, TSA+CTX—cefotaxime-resistant bacteria counts, TSA+OX—oxytetracycline-resistant bacteria counts, TSA+DOX—doxycycline—resistant bacteria counts, HRT— Hydraulic Retention Time, TSS—Total Suspended Solids, BOD—Biochemical Oxygen Demand, COD—Chemical Oxygen Demand.

3.2. Quantitative Analysis of Antibiotic Resistance Genes and Taxonomic Genes

3.2.1. Quantitative Pccurrence of Antibiotic Resistance Genes among the Total Number of Antibiotic-Resistant Bacteria

Among the analysed DNA samples of bacterial populations from the TSA universal culture medium, the presence of all the tested genes responsible for the resistance to beta-lactam antibiotics and tetracyclines was detected, with the exception of four wastewater treatment plants (No. VII, VIII, XI, X), in which the *bla*OXA gene was not detected (Figure 1).

The genes occurring at the highest average concentrations were *tet*A (from 8.72×10^{-4} to 1.07 gene copies/gene copies 16S *r*RNA) as well as *bla*OXA (from 8.54×10^{-3} to 4.53×10^{-1} gene copies/16S *r*RNA gene copies) and *bla*TEM (from 8.33×10^{-4} to 1.49×10^{-1} gene copies/16S *r*RNA gene copies). The other genes responsible for antibiotic resistance occurred at a similar level, at concentrations of the order of 1×10^{-3}–1×10^{-2} gene copies/16S *r*RNA gene copies. An analysis of the correlations between concentrations of particular genes originating from the total bacterial population revealed a statistically significant correlation between the *bla*SHV gene and all antibiotic resistance genes. Moreover, there was also a positive correlation between *bla*OXA and *tet*M, and between *bla*TEM and *tet*A genes (Table 1). An analysis of the correlation between the concentration of particular genes and the count of microorganisms demonstrated a statistically significant correlation between the concentration of the *bla*SHV gene in the DNA of the total bacterial population and the microorganism count on TSA media, both with and without the supplementation with the tested antibiotics. However, no significant correlation was found between physicochemical parameters and the concentration of genes, or between HRT values of the analysed treatment plants and the concentration of genes in wastewater.

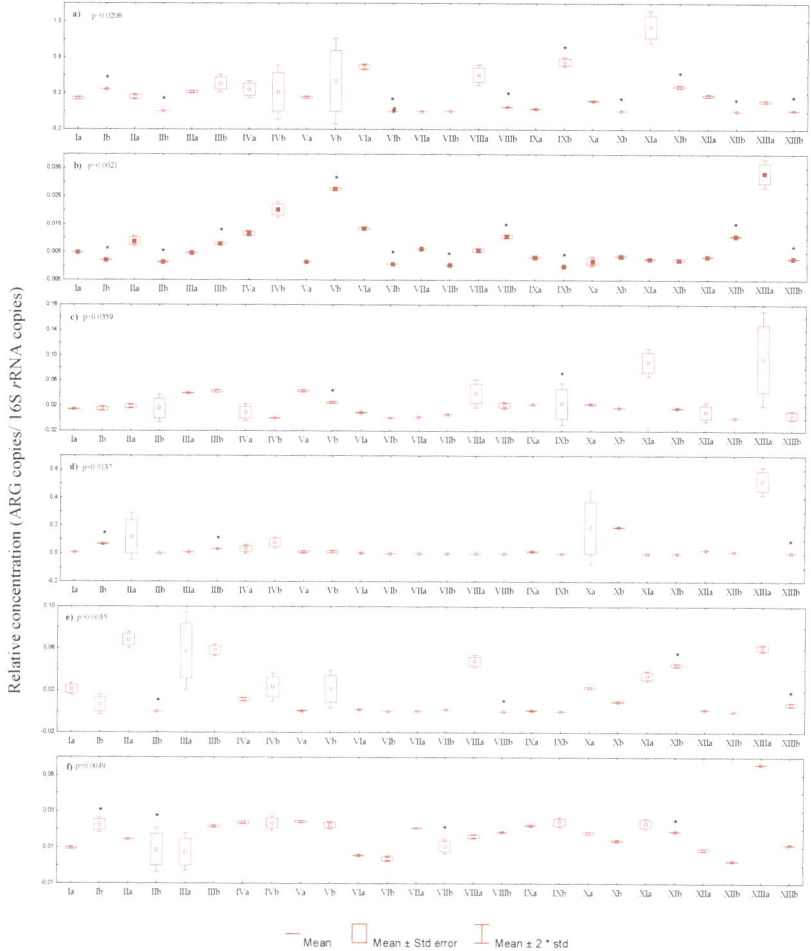

Figure 1. Relative antibiotic resistance gene (ARG) concentration in total bacteria population: (**a**) *tet*A (**b**) *tet*M (**c**) *bla*TEM (**d**) *bla*OXA (**e**) *bla*SHV (**f**) *uid*A in collected wastewater samples. An asterisk (*) denotes a statistically significant difference (ANOVA; $p < 0.05$) between wastewater treatment plant (WWTP) untreated and treated samples. I–XIII–denote numbers of WWTPs. Group A (I, II), group B (III–VII), group C (VIII–X), group D (XI–XIII); a—untreated wastewater, b—treated wastewater.

Following the wastewater treatment process, the average concentration of genes that are responsible for the resistance to beta-lactams and tetracyclines decreased at levels ranging from 0 to 100% (Table 2).

A reduction in the concentration of all antibiotic resistance genes under study was observed in the DNA of the total bacterial population from WWTP No. VI, group B (mechanical and biological treatment plants), and WWTP No. XIII, group D (increased nutrient removal mechanical and biological treatment plants). Moreover, in the DNA of the total bacterial population from other group D WWTPs, an increase was observed in the concentration of one gene in treated wastewater, i.e., *tet*M in WWTP No. XII and *bla*SHV in WWTP No. XI. Only in the DNA of the bacterial population from WWTP no. III, group B, was an increase noted in the concentrations of all the tested genes following the wastewater treatment process. However, in the DNA of the total bacterial population from WWTPS no. IV and V (group B), an increase was observed in the concentrations of four analysed genes (tetA, tetM, bla_{OXA}, and bla_{SHV}). Of all the drug resistance genes under study, the concentrations of bla_{SHV} and bla_{OXA} were reduced with the highest effectiveness, reaching as much as 100%. The genes that are responsible for the resistance to tetracyclines proved to be the most problematic group, since for both genes, tetA and tetM, an increase in their concentration in the TWW was observed in the DNA of the total bacterial population from six WWTPs.

The genes occurring at the highest average concentrations were tetA (from 3.57×10^{-2} to 1.94 gene copies/16S rRNA gene copies) and bla_{OXA} (from 8.21×10^{-3} to 6.63×10^{-1} gene copies/16S rRNA gene copies). The gene tetM was present at lower concentrations (from 9.18×10^{-8} to 4.29×10^{-1} gene copies/16S rRNA gene copies). For other genes encoding the resistance to beta-lactams, the average concentrations ranged from 1×10^{-3} to 1×10^{-2} and from 1×10^{-5} to 1×10^{-4} gene copies/16S rRNA gene copies, for the genes bla_{TEM} and bla_{SHV}, respectively. An analysis of the correlations between the concentrations of genes in the *E. coli* population DNA demonstrated a positive correlation between gene tetM and genes tetA, bla_{OXA}, and bla_{SHV}, and between gene bla_{TEM} and tetA (Table 3).

Table 2. Gene concentration reduction in total bacteria population.

Sewage Treatment Technology Used	Treatment Plant	tetA	tetM	bla_{TEM}	bla_{OXA}	bla_{SHV}	uidA
A. WWTPs with A$_2$O system	I	*	56.66	*	*	36.97	*
	II	99.88	83.58	*	100	100	*
B. WWTPs with mechanical-biological system	III	*	*	*	*	*	*
	IV	*	*	96.73	*	*	1.21
	V	*	*	42.45	*	*	7.97
	VI	99.87	94.27	93.90	100	97.05	31.93
	VII	*	93.87	*	**	*	49.51
C. WWTPs with Sequencing Batch Reactors (SBR)	VIII	87.89	*	46.94	**	100	*
	IX	*	99.98	*	100	90.66	*
	X	94.46	*	25.24	**	58.49	23.50
D. WWTPs with mechanical-biological system with elevated removal of nutrients	XI	70.34	13.17	81.96	**	*	19.44
	XII	98.68	*	95.82	47.98	100	72.85
	XIII	87.16	91.55	88.59	98.78	88.34	79.56

*—increase in gene concentration, **—no gene presence, WWTP—wastewater treatment plant.

3.2.2. Quantitative Occurrence of Antibiotic Resistance Genes among Antibiotic-Resistant *E. coli*

In the DNA of the bacterial population from *E. coli* selective medium mFc, tetM, tetA, and bla_{OXA} genes were not detected in untreated wastewater from all treatment plants (Figure 2).

Following the wastewater treatment process, the average concentration of genes that are responsible for the resistance to beta-lactams and tetracyclines decreased at a level ranging from 0 to 100% (Table 4).

A complete reduction in genes was observed for the tetA gene in the DNA of *E. coli* population from four wastewater treatment plants (No. VII, VIII, X, XI) and for the bla_{OXA} gene for the DNA of *E. coli* population isolated from WWTP No. III. A reduction in all genes under study was observed only in the DNA of *E. coli* populations from WWTPs No. VIII and IX, group C, ranging from 21% to 100%. An increase in the concentration of all genes under study following the wastewater treatment process was observed in the DNA of *E. coli* population from WWTP No. IV, group B. In addition, an increase in the concentrations of four genes under study (tetA, tetM, bla_{OXA}, bla_{TEM}) was also observed in the DNA of *E. coli* population in WWTP No. V, group B, and gene bla_{SHV} was reduced only by 8%. Wastewater treatment plants reduced the concentration of the bla_{OXA} gene most efficiently. However, the bla_{TEM} gene, whose concentration increased following the treatment process, was noted in the

DNA of the population of *E. coli* from as many as eight WWTPs and was the most difficult to remove from wastewater.

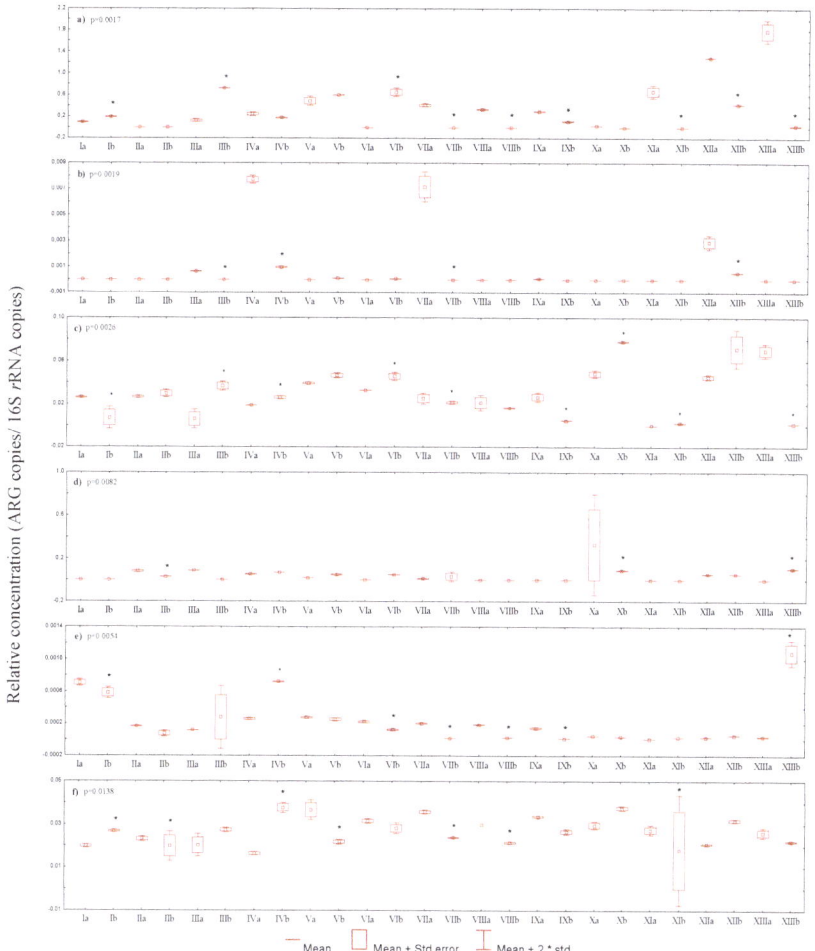

Figure 2. Relative ARGs concentration in *E. coli* population: (**a**) *tet*A (**b**) *tet*M (**c**) *bla*TEM (**d**) *bla*OXA (**e**) *bla*SHV (**f**) *uid*A in collected wastewater samples. An asterisk (*) denotes a statistically significant difference (ANOVA; $p < 0.05$) between WWTP influent and effluent samples. I-XIII—denote numbers of WWTPs. Group A (I, II), group B (III–VII), group C (VIII–X), group D (XI–XIII); a—untreated wastewater, b—treated wastewater.

Table 3. Correlations between gene concentrations, *E. coli* counts, WWTP parameters, and basic physico-chemical parameters.

	mFc/ 1 mL	mFc-AMO/ 1 mL	mFc-CTX/ 1 mL	mFc-OX/ 1 mL	mFc-DOX/ 1 mL	*tet*A	*tet*M	*bla*TEM	*bla*OXA	*bla*SHV	*uid*A	HRT	TSS	BOD
								Spearman's Rank Coeffcient						
mFc-AMO/ 1 mL	0.93													
mFc-CTX/ 1 mL	0.66	0.74												
mFc-OX/ 1 mL	0.81	0.83	0.59											
mFc-DOX/ 1 mL	0.68	0.69	0.65	0.73										
*tet*A	0.24	0.13	0.13	0.14	0.07									
*tet*M	0.17	0.14	0.17	0.05	−0.02	0.56								
*bla*TEM	0.16	0.07	−0.11	0.09	0.13	**0.30**	0.24							
*bla*OXA	−0.09	−0.02	−0.12	0.02	−0.03	−0.12	0.28	0.22						
*bla*SHV	0.10	0.13	0.15	0.01	0.00	0.05	0.37	−0.02	0.21					
*uid*A	−0.04	−0.12	−0.32	−0.15	−0.06	0.11	0.06	**0.32**	−0.04	0.02				
HRT	−0.05	−0.08	0.08	−0.11	−0.23	0.03	0.14	**−0.61**	−0.38	−0.10	−0.12			
TSS	−0.01	0.06	**0.46**	0.14	0.23	−0.03	−0.31	−0.05	−0.12	−0.19	−0.01	−0.07		
BOD	**−0.63**	**−0.63**	**−0.69**	**−0.54**	**−0.43**	−0.31	−0.35	−0.11	−0.13	−0.23	0.05	0.04	−0.15	
COD	−0.29	−0.38	−0.28	**−0.41**	−0.34	0.20	0.27	0.17	−0.21	−0.09	0.25	0.19	−0.28	0.12

mFc—total *E. coli* counts, mFc+AMO—amoxicillin-resistant *E. coli* counts, mFc+CTX—cefotaxime-resistant *E. coli* counts, mFc+OX—oxytetracycline-resistant *E. coli* counts, mFc+DOX—doxycycline-resistant *E. coli* counts, HRT—Hydraulic Retention Time, TSS—Total Suspended Solids, BOD—Biochemical Oxygen Demand, COD—Chemical Oxygen Demand.

Table 4. Gene concentration reduction in *E. coli* populations.

Sewage Treatment Technology Used	Treatment Plant	*tet*A	*tet*M	*bla*TEM	*bla*OXA	*bla*SHV	*uid*A
A. WWTPs with A₂O system	I	*	93.66	44.30	**	17.50	*
	II	**	**	*	64.30	53.95	14.67
B. WWTPs with mechanical-biological system	III	*	99.66	*	100	*	*
	IV	*	*	*	*	*	*
	V	*	*	*	*	7.63	40.10
	VI	**	*	*	**	43.52	10.82
	VII	100	99.997	14.50	*	90.47	33.22
C. WWTPs with Sequencing Batch Reactors (SBR)	VIII	100	86.13	21.88	**	84.69	27.77
	IX	60.98	99.76	80.98	**	88.40	20.75
	X	100	68.71	*	85.93	26.06	*
D. WWTPs with mechanical-biological system with elevated removal of nutrients	XI	100	*	*	**	*	*
	XII	66.62	80.96	*	*	*	*
	XIII	98.21	19.55	97.30	**	*	15.52

*—increase in gene concentration, **—no gene presence, WWTP—wastewater treatment plant.

3.2.3. Quantitative Occurrence of *uid*A Genes

The average concentration of the taxonomic gene of *E. coli* ranged from 9.97×10^{-3} to 5.63×10^{-2} gene copies/16S *r*RNA gene copies for the DNA from the total population of bacteria isolated on TSA medium, and from 1.50×10^{-2} to 4.29×10^{-1} gene copies/16S *r*RNA gene copies for the DNA derived from the population of *E. coli* isolated on the mFc medium. The concentrations of genes *uid*A in the DNA of the total bacterial population were positively correlated with the concentration of genes *tet*A and *bla*TEM, while concentrations of genes *uid*A in the DNA of *E. coli* population were only positively correlated with gene *bla*TEM. The greatest reduction in the concentration of *E. coli*, based on the taxonomic gene, was observed in WWTP No. XII, group D (80%) in the DNA of the total bacterial population and in WWTP No. V, group B (40%) for the DNA of *E. coli* population. The relative concentration of gene *uid*A following the wastewater treatment process increased for the DNA obtained from *E. coli* population in six wastewater treatment plants, while in the DNA from the total bacterial population, from five WWTPs.

4. Discussion

Wastewater treatment plants constitute a potential source of ARB and ARGs, which can be transmitted to the environment with discharged treated wastewater. For this reason, monitoring microbiological pollutant removal effectiveness during wastewater treatment is of particular significance. Results of studies on the number of antibiotic resistant bacteria on culture media were published by Osińska et al. [29]. Based on these results, the most frequently occurring bacteria were those that were resistant to amoxicillin and oxytetracycline, i.e., older generation antibiotics. Even though wastewater treatment plants usually ensured a rather high level of reduction in the number of antibiotic-resistant bacteria (at least 99.9%), treated wastewater from WWTPs that were discharged to the environment contained, on average, from 1.05×10^2 to 2.77×10^5 CFU/mL. It should be stressed that the percentage of antibiotic-resistant bacteria in the total number of microorganisms increased following the wastewater treatment process.

Genes that are responsible for resistance to beta-lactam antibiotics were reduced to a greater extent following the wastewater treatment process as compared to the number of genes responsible for resistance to tetracyclines. Czekalski et al. [6] also found that, despite a reduction by as much as 78% in the total number of bacteria following the wastewater treatment process, a reduction in ARGs concentration was not always observed. Moreover, the occurrence of multi-drug resistant bacteria was found. Rafrat et al. [31] demonstrated that the concentration of $bla_{\text{CTX-M}}$, bla_{TEM}, and qnrS genes relative to the number of copies of 16S *rRNA* was generally similar in the samples of wastewater both before and after the treatment process. The authors also demonstrated that in wastewater from the treatment plant receiving mixed municipal and hospital wastewater, the concentration of genes increased following the treatment process. In wastewater from a biological treatment plant using disinfection with UV radiation, there was also no reduction in the concentrations of bla_{TEM}, qnrA, and sulI genes in the samples of treated wastewater from that WWTP. Rodriguez-Mozaz et al. [23] noted that the relative concentrations of ermB and tetW genes decreased as a result of wastewater treatment, while the concentration of bla_{TEM}, sul1, and qnrS genes in wastewater increased. Those authors also demonstrated that the incomplete removal of antibiotics and ARGs in the wastewater treatment process significantly affected the water quality of the river receiving treated wastewater. This was because the concentration of antibiotics and ARG concentrations at a point beyond the wastewater inflow were higher than in the samples collected before the effluent discharge point. Other study results [24,25] also confirm the ineffective elimination of ARGs by conventional wastewater treatment plants. They also indicate that the bacteria released with wastewater from WWTPs may be capable of actively spreading the resistance genes among the local microorganisms, thus inhabiting ecosystems of surface water bodies that receive treated wastewater.

The present study demonstrated that for the genes responsible for resistance to tetracyclines in all tested WWTPs, both in the total bacterial population and in *E. coli* population, the *tet*A gene occurred at higher concentrations than the *tet*M gene. Zhang et al. [18]. in their study concerning the change in microorganism communities of the activated sludge and ARGs under selection pressure of antibiotics in SBR reactors, observed that the total number of ARGs in the activated sludge increased following the addition of tetracycline (TC). This indicates that bacteria with ARGs proliferate intensively under the pressure of antibiotic presence. Moreover, the study demonstrated that, with an increase in TC concentration, the concentration of genes that are mainly responsible for the efflux pump (*tet*A, *tet*B, *tet*C) increased, while the concentration of genes responsible for the abundance of ribosomal genes of protective proteins (*tet*M, *tet*O, *tet*S) was lower when compared with the number of the efflux pump genes and that this number changed slightly with an increase in TC concentration.

The present study also noted the highest concentrations of *tet*A and bla_{OXA} genes in the DNA of *E. coli* populations, while the occurrence of *tet*M and bla_{OXA} genes was not noted in all the studied samples of the DNA of the *E. coli* populations from wastewater treatment plants. As a result of the intake of antibiotics, antibiotic resistance genes can be modified and acquired by the microbiota inhabiting the gastrointestinal tract, which may be released to the environment with faeces via drainage systems [23]. Osińska et al. [28] found that, following the wastewater treatment process, the frequency of the occurrence of antibiotic resistance genes increased among *E. coli* bacteria. The authors also noted an increase in the percentage of multi-drug resistant *E. coli* as compared to untreated wastewater. Hendricks and Pool [32] suggest that wastewater treatment plants are ineffective in eliminating bacteria, including faecal contamination indicator bacteria, even though studies with opposite results have been reported [30]. These great differences in the efficiency of processed wastewater treatment may result from the wastewater treatment plant capacity and/or the type of wastewater [33]. Hembach et al. [4] also found a slight reduction in the concentration of ARGs during standard wastewater treatment at seven wastewater treatment plants of different capacities.

The conducted study indicated concentrations of *uid*A gene that enable the determination of *E. coli* of an order up to 10^{-1} copies/16S *r*RNA gene copies. In most cases, no significant reductions in the number of copies of this gene following the wastewater treatment process were noted. Rafrat et al. [31] demonstrated that the number of copies of bacterial 16S *r*RNA was up to 5.84×10^8 copies/mL of sample, and there were slight differences ($p > 0.05$) between the number of gene copies in untreated and treated wastewater. The results of the studies of those authors suggest that the applied wastewater treatment processes were not effective in reducing the bacterial load, which is also true for wastewater treatment plants applying UV disinfection following the conventional treatment process. Even though a study by Hendricks and Pool [32] demonstrated that wastewater treatment plants may decrease the load of nutrients and faecal coliforms from treated wastewater, Rafrat et al. [31] indicated that the total abundance of bacteria (estimated based on the number of copies of the bacterial gene 16S rRNA) was not reduced following the treatment process, irrespective of the process applied. Therefore, biological wastewater treatment plants may provide an appropriate environment for the spread of ARGs, as both the conditions prevailing during wastewater treatment, and the great diversity of microorganisms facilitate the transmission of genes through, *inter alia*, horizontal gene transfer (HGT) [6].

For most WWTPs receiving only municipal wastewater, the *bla*$_{OXA}$ gene was not detected in the DNA obtained from the total population of bacteria isolated on TSA medium or in the DNA of the *E. coli* population. In the DNA of populations that were isolated from wastewater in other treatment plants, the concentration of this gene was dominant. Another quantitatively dominant gene responsible for resistance to beta-lactams was the *bla*$_{TEM}$ gene. It is believed that the group of TEM-type beta-lactamases may be of paramount significance in the development of resistance to beta-lactam antibiotics due to its great diversity (over 220 types of genes belong to this group) [34]. The results that were obtained by Ojer-Usoz et al. [35] also indicated the dominant occurrence of *bla*$_{TEM}$ gene in relation to the *bla*$_{SHV}$ and *bla*$_{OXA}$ genes in the DNA of isolates originating from wastewater treatment plants. However, this study found that WWTPs that received only municipal wastewater proved to exhibit a much better performance in reducing the analysed genes responsible for antibiotic resistance than wastewater treatment plants also receiving wastewater from the food industry and hospitals.

Following the wastewater treatment process, an increase or a slight reduction in the concentration of studied ARGs was observed in the DNA of the total bacterial population and in the DNA of the *E. coli* population, mainly in WWTPs with a mechanical and biological system (group B). Moreover, an increase in the concentration of all the studied genes was observed in the DNA of the total bacterial population from WWTP No. II and in the DNA of the *E. coli* population isolated from WWTP No. IV. An increase in the ARG concentration following the wastewater treatment process was also found in the DNA samples from both studied populations isolated from WWTP No. V with a mechanical and biological system. Moreover, an over 90% decrease in all ARGs in the DNA of the total bacterial population was noted in WWTP No. VI, which used a mechanical and biological wastewater treatment system. At the same time, an increase in the number of *tet*M and *bla*$_{TEM}$ gene copies and the presence of *tet*A gene was observed in the DNA of the *E. coli* populations that were isolated from the same plant. Interestingly enough, the latter gene was not detected in the untreated wastewater received by the plant. The highest reduction in the studied ARGs was recorded for the WWTP No. XII (group D) with an increased nutrient removal mechanical and biological system in the DNA samples isolated from the total bacterial populations and for the WWTPs No. VIII and IX with sequencing batch reactors (group C) in the DNA samples that were obtained from the *E. coli* populations. Rafrat et al. [31] also found differences in the concentrations of genes in treated wastewater from various WWTPs, including an increase in the concentration of genes in wastewater from a treatment plant which receives municipal and hospital wastewater. The differences in ARG concentrations between wastewater samples from various treatment plants may be due to the differences in microbiological composition of wastewater incoming to particular treatment plants and the diversity of antibiotics (concentration and type) present in wastewater [17]. Guo et al. [36] found that ARGs from chromosomal mutations, under the great selective pressure of antibiotics, decreased along with the degradation of total antibiotics in final effluents, instead of proliferating through biological treatment stages. Youan et al. [35] demonstrated that the ARB count and the ARG number were statistically significantly positively correlated with the COD and the contaminant load, and negatively correlated with the oxygen content and the wastewater temperature. Moreover, Jiao et al. [37] observed that the absolute concentration of ARGs decreased by one order of magnitude in the DNA of samples that were collected in winter as compared with the DNA samples collected in summer, which could be explained by the fact that low temperatures may cause microbial enzyme deactivation, thus having a direct effect on ARG removal effectiveness. These correlations were not confirmed by Wen et al. [25], who suggest that varied temperatures throughout different seasons does not have a significant effect on the ARG concentration. These differences may be explained by regional diversification of antibiotic doses or by other factors that are connected with seasonal temperature fluctuations. A factor with a significant effect on the ARG concentrations may also be the composition of the wastewater reaching WWTPs. Numerous studies have documented high concentrations of ARGs in wastewater produced by hospitals [6,23,30] or some industries, e.g., the food industry [11,20,38]. Moreover, dissemination of ARGs may also be affected by wastewater contamination with heavy metals, whose residues may pose co-selection pressure on ARG distribution [24]. For this reason, ARGs are not effectively eliminated in the wastewater treatment process and their concentration may even increase [39]. Wastewater discharge to the environment or their re-utilisation may lead to the dissemination of ARB and ARGs in the environment, threatening the ecological safety of aquatic ecosystems, in particular [27,40]. Moreover, the continued ARB and ARG discharge to the environment may contribute to an increase in infections with resistant pathogens and may enlarge the ARG pool among the environmental bacteria [6]. At the same time, Lorenzo et al. [41] demonstrated that, despite an effective reduction in absolute ARB and ARG concentrations by wastewater treatment plants, they might still remain hot-spots of antibiotic resistance expansion among wastewater bacteria sensitive to these drugs and considerably increase the antibiotic resistance levels in freshwater ecosystems that receive such wastewater. The results of the present study of the role of wastewater treatment plants in antibiotic resistance have demonstrated that the ARG concentrations in wastewater discharged from WWTPs may fluctuate significantly and the factors determining ARG elimination effectiveness in the wastewater treatment process should be monitored.

5. Conclusions

The results of the present study indicate that WWTPs play a significant role in the acquisition and spread of antibiotic resistance genes among bacterial populations. The presence of ARGs in live bacterial cells that are capable of multiplying constitutes a considerably greater potential threat of transmission of these genes from WWTPs to the environment. Therefore, the present study may be helpful in identifying the risks to human public health and the well-being of animals exploiting water bodies that receive treated wastewater. The analysed genes of resistance to beta-lactam antibiotics (*bla*$_{OXA}$, *bla*$_{TEM}$, *blu*$_{SHV}$) and to tetracyclines (*tet*A, *tet*M) were detected in the DNA of bacterial populations isolated from wastewater from most wastewater treatment plants under study. In the DNA of *E. coli* population, the occurrence of *tet*M and *bla*$_{OXA}$ genes was not detected in any of the tested WWTPs. It should be noted that not all gene concentrations were reduced following the treatment processes. For certain genes (*bla*$_{TEM}$, *tet*A), their average concentrations increased by as much as one order of magnitude following the treatment process. The results of this study indicate the presence of significant concentrations of genes that are responsible for antibiotic resistance, even in the treated wastewater from WWTPs. Generally, in wastewater treatment plants with a mechanical-biological system, an increased number of ARGs copies after treatment processes were observed. This indicates the need for continuous monitoring of wastewater treatment processes, as the discharge of wastewater to aquatic environments following treatment processes may pose a serious hazard to environmental safety.

Supplementary Materials: The following are available online at http://www.mdpi.com/2076-3417/9/3/387/s1, Table S1: Technical description of wastewater treatment plants (WWTPs) and physicochemical parameters of wastewater, Table S2: Quantitative PCR (qPCR) primers and conditions used in this study, Table S3: Correlations between concentration of genes, total number of bacteria, WWTP parameters and basic physico-chemical parameters, Table S4: Correlations between concentration of genes, number of *E. coli*, WWTP parameters and basic physico-chemical parameters.

Author Contributions: Data curation, E.K. and A.O.; Investigation, E.K., M.H. and A.O.; Methodology, E.K. and M.H. and A.O.; Writing—original draft, A.O. and E.K.; Writing—review & editing, A.O., E.K., M.H. and S.N.

Funding: This study was supported by grants No. 2017/27/B/NZ9/00267 and No. 2016/23/BNZ9/03669 from the National Science Center (Poland).

Acknowledgments: We would like to thank the staff of the WWTPs for the possibility of collecting samples.

Conflicts of Interest: The authors declare no conflict of interest.

References

1. Leonard, A.F.C.; Zhang, L.H.; Balfour, A.; Garside, R.; Gaze, W.H. Human recreational exposure to antibiotic resistant bacteria in coastal bathing waters. *Environ. Int.* **2015**, *82*, 92–100. [CrossRef] [PubMed]

2. World Health Organization. Antimicrobial Resistance: Global Report on Surveillance 2014. Available online: http://www.who.int/en/ (accessed on 6 February 2017).

3. Martinez, J.L.; Coque, T.M.; Baquero, F. What is a resistance gene? Ranking risk in resistomes. *Nat. Rev. Microbiol.* **2015**, *13*, 116–123. [CrossRef] [PubMed]

4. Hembach, N.; Schmid, F.; Alexander, J.; Hiller, C.; Rogall, E.T.; Schwartz, T. Occurrence of the mcr-1 Colistin Resistance Gene and other Clinically Relevant Antibiotic Resistance Genes in Microbial Populations at Different Municipal Wastewater Treatment Plants in Germany. *Front. Microbiol.* **2017**, *8*, 1282. [CrossRef] [PubMed]

5. Alexander, J.; Bollmann, A.; Seitz, W.; Schwartz, T. Microbiological characterization of aquatic microbiomes targeting taxonomical marker genes and antibiotic resistance genes of opportunistic bacteria. *Sci. Total Environ.* **2015**, *512*, 316–325. [CrossRef] [PubMed]

6. Czekalski, N.; Berthold, T.; Caucci, S.; Egli, A.; Burgmann, H. Increased levels of multiresistant bacteria and resistance genes after wastewater treatment and their dissemination into Lake Geneva, Switzerland. *Front. Microbiol.* **2012**, *3*, 106. [CrossRef]

7. Hsu, C.Y.; Hsu, B.M.; Ji, W.T.; Chen, J.S.; Hsu, T.K.; Ji, D.D.; Tseng, S.F.; Chiu, Y.C.; Kao, P.M.; Huang, Y.L. Antibiotic resistance pattern and gene expression of non-typhoid Salmonella in riversheds. *Environ. Sci. Pollut. Res.* **2015**, *22*, 7843–7850. [CrossRef]

8. Gillings, M.R.; Gaze, W.H.; Pruden, A.; Smalla, K.; Tiedje, J.M.; Zhu, Y.G. Using the class 1 integron-integrase gene as a proxy for anthropogenic pollution. *ISME J.* **2015**, *9*, 1269–1279. [CrossRef]

9. Martinez, J.L. Antibiotics and antibiotic resistance genes in natural environments. *Science* **2008**, *321*, 365–367. [CrossRef]

10. Salyers, A.; Shoemaker, N.B. Reservoirs of antibiotic resistance genes. *Anim. Biotechnol.* **2006**, *17*, 137–146. [CrossRef]

11. Harnisz, M.; Korzeniewska, E.; Gołaś, I. The impact of a freshwater fish farm on the community of tetracycline-resistant bacteria and the structure of tetracycline resistance genes in river water. *Chemosphere* **2015**, *128*, 134–141. [CrossRef]

12. Kotlarska, E.; Łuczkiewicz, A.; Pisowacka, M.; Burzyński, A. Antibiotic resistance and prevalence of class 1 and 2 integrons in Escherichia coli isolated from two wastewater treatment plants, and their receiving waters (Gulf of Gdansk, Baltic Sea, Poland). *Environ. Sci. Pollut. Res.* **2015**, *22*, 2018–2030. [CrossRef] [PubMed]

13. Pruden, A. Balancing Water Sustainability and Public Health Goals in the Face of Growing Concerns about Antibiotic Resistance. *Environ. Sci. Technol.* **2014**, *48*, 5–14. [CrossRef] [PubMed]

14. Harnisz, M.; Korzeniewska, E. The prevalence of multidrug-resistant *Aeromonas* spp. in the municipal wastewater system and their dissemination in the environment. *Sci. Total Environ.* **2018**, *626*, 377–383. [CrossRef] [PubMed]

15. Berglund, B.; Fick, J.; Lindgren, P.E. Urban wastewater effluent increases antibiotic resistance gene concentrations in a receiving northern European river. *Environ. Toxicol. Chem.* **2015**, *34*, 192–196. [CrossRef]

16. Rizzo, L.; Manaia, C.; Merlin, C.; Schwartz, T.; Dagot, C.; Ploy, M.C.; Michael, I.; Fatta-Kassinos, D. Urban wastewater treatment plants as hotspots for antibiotic resistant bacteria and genes spread into the environment: A review. *Sci. Total Environ.* **2013**, *447*, 345–360. [CrossRef] [PubMed]

17. Yang, Y.; Li, B.; Zou, S.C.; Fang, H.H.P.; Zhang, T. Fate of antibiotic resistance genes in sewage treatment plant revealed by metagenomic approach. *Water Res.* **2014**, *62*, 97–106. [CrossRef] [PubMed]

18. Zhang, Y.Y.; Geng, J.J.; Ma, H.J.; Ren, H.Q.; Xu, K.; Ding, L.L. Characterization of microbial community and antibiotic resistance genes in activated sludge under tetracycline andsulfamethoxazole selection pressure. *Sci. Total Environ.* **2016**, *571*, 479–486. [CrossRef]

19. Biswal, B.K.; Mazza, A.; Masson, L.; Gehr, R.; Frigon, D. Impact of wastewater treatment processes on antimicrobial resistance genes and their co-occurrence with virulence genes in *Escherichia coli*. *Water Res.* **2014**, *50*, 245–253. [CrossRef]

20. Huang, K.L.; Tang, J.Y.; Zhang, X.X.; Xu, K.; Ren, H.Q. A Comprehensive Insight into Tetracycline Resistant Bacteria and Antibiotic Resistance Genes in Activated Sludge Using Next-Generation Sequencing. *Int. J. Mol. Sci.* **2014**, *15*, 10083–10100. [CrossRef]

21. Davies, J. Are antibiotics naturally antibiotics? *J. Ind. Microbiol. Biotechnol.* **2006**, *33*, 496–499. [CrossRef]

22. Warnes, S.L.; Highmore, C.J.; Keevil, C.W. Horizontal Transfer of Antibiotic Resistance Genes on Abiotic Touch Surfaces: Implications for Public Health. *mBio* **2012**, *3*, e00489-12. [CrossRef] [PubMed]

23. Rodriguez-Mozaz, S.; Chamorro, S.; Marti, E.; Huerta, B.; Gros, M.; Sanchez-Melsio, A.; Borrego, C.M.; Barceló, D.; Balcázar, J.L. Occurrence of antibiotics and antibiotic resistance genes in hospital and urban wastewaters and their impact on the receiving river. *Water Res.* **2015**, *69*, 234–242. [CrossRef] [PubMed]

24. Di Cesare, A.; Eckert, E.M.; D'Urso, S.; Bertoni, R.; Gillan, D.C.; Wattiez, R.; Corno, G. Co-occurrence of integrase 1, antibiotic and heavy metal resistance genes in municipal wastewater treatment plants. *Water Res.* **2016**, *94*, 208–214. [CrossRef] [PubMed]

25. Wen, Q.X.; Yang, L.; Duan, R.; Chen, Z.Q. Monitoring and evaluation of antibiotic resistance genes in four municipal wastewater treatment plants in Harbin, Northeast China. *Environ. Pollut.* **2016**, *212*, 34–40. [CrossRef] [PubMed]

26. Bengtsson-Palme, J.; Hammarén, R.; Pal, C.; Östman, M.; Björlenius, B.; Flach, C.F.; Fick, J.; Kristiansson, E.; Tysklind, M.; Larsson, D.G.J. Elucidating selection processes for antibiotic resistance in sewage treatment plants using metagenomics. *Sci. Total Environ.* **2016**, *572*, 697–712. [CrossRef] [PubMed]

27. Ben, W.W.; Wang, J.; Cao, R.K.; Yang, M.; Zhang, Y.; Qiang, Z.M. Distribution of antibiotic resistance in the effluents of ten municipal wastewater treatment plants in China and the effect of treatment processes. *Chemosphere* **2017**, *172*, 392–398. [CrossRef] [PubMed]

28. Osińska, A.; Korzeniewska, E.; Harnisz, M.; Niestępski, S. The prevalence and characterization of antibiotic-resistant and virulent Escherichia coli strains in the municipal wastewater system and their environmental fate. *Sci. Total Environ.* **2017**, *577*, 367–375. [CrossRef]

29. Osińska, A.; Korzeniewska, E.; Harnisz, M.; Niestępski, S. Impact of type of wastewater treatment process on the antybiotic resistance of bacterial populations. *E3S Web Conf.* **2017**, *17*, 00070. [CrossRef]

30. Korzeniewska, E.; Harnisz, M. Extended-spectrum beta-lactamase (ESBL)-positive Enterobacteriaceae in municipal sewage and their emission to the environment. *J. Environ. Manag.* **2013**, *128*, 904–911. [CrossRef]

31. Rafraf, I.D.; Lekunberri, I.; Sanchez-Melsio, A.; Aouni, M.; Borrego, C.M.; Balcazar, J.L. Abundance of antibiotic resistance genes in five municipal wastewater treatment plants in the Monastir Governorate, Tunisia. *Environ. Pollut.* **2016**, *219*, 353–358. [CrossRef]

32. Hendricks, R.; Pool, E.J. The effectiveness of sewage treatment processes to remove faecal pathogens and antibiotic residues. *J. Environ. Sci. Health A* **2012**, *47*, 289–297. [CrossRef] [PubMed]

33. Koivunen, J.; Siitonen, A.; Heinonen-Tanski, H. Elimination of enteric bacteria in biological-chemical wastewater treatment and tertiary filtration units. *Water Res.* **2003**, *37*, 690–698. [CrossRef]

34. Jacoby, G.A.; Munoz-Price, L.S. The new beta-lactamases. *N. Engl. J. Med.* **2005**, *352*, 380–391. [CrossRef]

35. Ojer-Usoz, E.; Gonzalez, D.; Garcia-Jalon, I.; Vitas, A.I. High dissemination of extended-spectrum beta-lactamase-producing Enterobacteriaceae in effluents from wastewater treatment plants. *Water Res.* **2014**, *56*, 37–47. [CrossRef]

36. Guo, X.; Yan, Z.; Zhang, Y.; Xu, W.; Kong, D.; Shan, Z.; Wang, N. Behavior of antibiotic resistance genes under extremely high-level antibiotic selection pressures in pharmaceutical wastewater treatment plants. *Sci. Total Environ.* **2018**, *612*, 119–128. [CrossRef] [PubMed]

37. Jiao, Y.N.; Zhou, Z.C.; Chen, T.; Wei, Y.Y.; Zheng, J.; Gao, R.X. Biomarkers of antibiotic of resistance genes during seasonal changes in wastewater treatment systems. *Environ. Pollut.* **2018**, *234*, 79–87. [CrossRef] [PubMed]

38. Yuan, Q.B.; Guo, M.T.; Yang, J. Monitoring and assessing the impact of wastewater treatment on release of both antibiotic-resistant bacteria and their typical genes in a Chinese municipal wastewater treatment plant. *Environ. Sci.* **2014**, *16*, 1930–1937. [CrossRef] [PubMed]

39. Zanotto, C.; Bissa, M.; Illiano, E.; Mezzanotte, V.; Marazzi, F.; Turolla, A.; Antonelli, M.; De Giuli Morghen, C.; Radaelli, A. Identification of antibiotic-resistant Escherichia coli isolated from a municipal wastewater treatment plant. *Chemosphere* **2016**, *164*, 627–633. [CrossRef] [PubMed]

40. Makowska, N.; Koczura, R.; Mokracka, J. Class 1 integrase, sulfonamide and tetracycline resistance genes in wastewater treatment plant and surface water. *Chemosphere* **2016**, *144*, 1665–1673. [CrossRef] [PubMed]

41. Lorenzo, P.; Adriana, A.; Jessica, S.; Carles, B.; Marinella, F.; Marta, L.; Pierre, S. Antibiotic resistance in urban and hospital wastewaters and their impact on a receiving freshwater ecosystem. *Chemosphere* **2018**, *206*, 70–82. [CrossRef]

Article

Environmental Pollutants Impair Transcriptional Regulation of the Vitellogenin Gene in the Burrowing Mud Crab (*Macrophthalmus Japonicus*)

Kiyun Park [1], Hyunbin Jo [1], Dong-Kyun Kim [1] and Ihn-Sil Kwak [1,2,*]

[1] Fisheries Science Institute, Chonnam National University, Yeosu 59626, Korea; ecoblue@hotmail.com (K.P.); prozeva@hanmail.net (H.J.); dkkim1004@gmail.com (D.-K.K.)
[2] Faculty of Marine Technology, Chonnam National University, Yeosu 550-749, Korea
* Correspondence: iskwak@chonnam.ac.kr; Tel.: +82-61-6597148; Fax: +82-61-6597149

Received: 14 February 2019; Accepted: 1 April 2019; Published: 3 April 2019

Abstract: Vitellogenesis is a pivotal reproductive process of the yolk formation in crustaceans. Vitellogenin (VTG) is the precursor of main yolk proteins and synthesized by endogenous estrogens. The intertidal mud crab (*Macrophthalmus japonicus*) inhabits sediment and is a good indicator for assessing polluted benthic environments. The purpose of this study was to identify potential responses of *M. japonicus* VTG under environmental stresses caused by chemical pollutants, such as 1, 10, and 30 µg L^{-1} concentrations in di(2-ethylhexyl) phthalate (DEHP), bisphenol A (BPA) and irgarol. We characterized the *M. japonicus* VTG gene and analyzed the transcriptional expression of VTG mRNA in *M. japonicus* exposed to various chemicals and exposure periods. A phylogenetic analysis revealed that the *M. japonicus* VTG clustered closely with *Eriocheir sinensis* (Chinese mitten crab) VTG, in contrast with another clade that included the VTG ortholog of other crabs. The basal level of VTG expression was the highest in the hepatopancreas and ovaries, and tissues. VTG expression significantly increased in the ovaries and hepatopancreas after 24 h exposure to DEHP. Increased responses of VTG transcripts were found in *M. japonicus* exposed to DEHP and BPA for 96 h; however, VTG expression decreased in both tissues after irgarol exposure. After an exposure of 7 d, VTG expression significantly increased in the ovaries and hepatopancreas for all concentrations of all chemicals. These results suggest that the crustacean embryogenesis and endocrine processes are impaired by the environmental chemical pollutants DEHP, BPA, and irgarol.

Keywords: vitellogenin (VTG); crustacean; di(2-ethylhexyl) phthalate (DEHP); bisphenol A (BPA); irgarol

---

1. Introduction

Vitellogenesis is a hormonally regulated process for synthesizing yolk proteins and an important step in the reproductive development of crustaceans [1]. Vitellogenin (VTG), a precursor molecule of vitellin, is stored in crustacean hepatopancreas tissue and transported by the hemolymph to the ovaries. The sites of VTG synthesis are in the hepatopancreas and ovary tissues in decapod crustaceans [2–5]. VTG genes have shown specific levels of expressions that vary by tissue, sex, and development stage [6]. VTG inductions were observed in aquatic invertebrates by exposure to estrogenic compounds [7,8]. After endosulfan exposure, VTG gene expression was downregulated in Pandalus shrimp (*Pandalopsis japonica*) [9]. The regulation of VTG gene may be affected by several pollutants [7–9].

Environmental chemical pollution is globally one of the most critical ecological problems, owing to the high risks it poses to ecosystems and human life [10]. Bisphenol A (BPA), a common endocrine disrupting chemical (EDC), is a carbon-based synthetic compound and a xenoestrogen that interferes with estrogen receptor signaling [11]. BPA concentrations observed in U.S. river estuaries were approximately 4 μg L^{-1} to 21 μg L^{-1} [12,13]. Exposure to BPA induced transcription responses of the heat shock protein 90 (HSP90) gene in the marine crab *Charybdis japonica* [14]. Di-(2-ethylhexyl) phthalate (DEHP), a suspected EDC, is widely used as plasticizer and may induce reproductive and developmental toxicities in aquatic environments [15]. DEHP toxicity induced changes in the VTG levels in African sharptooth catfish (*Clarias gariepinus*) and the Chinese rare minnow (*Gobiocypris rarus*) [16,17]. DEHP levels were 0.47 to 12 μg L^{-1} in the China river estuary [18]. Irgarol, which is used as an algicide, has been found in coastal and estuarine environments because it is commonly used in antifouling systems [19,20]. Irgarol exposure induced downregulation of HSP90 expression in hard coral (*Acropora tenuis*) [21]. Worldwide, the environmental concentrations of irgarol in antifouling systems range from 1 ng L^{-1} to 1 mg L^{-1} [22]. Irgarol's long half-life poses an ecological risk in estuarine ecosystems [20]. Although these pollutants are ubiquitously distributed in aquatic environments, there is not enough information about the effects of embryogenesis in crustaceans.

Crabs make up one of the largest crustacean families. The accumulation and synthesis of VTG is vital for oocyte and embryo development in crabs [2,6,23]. A burrowing mud crab, *Macrophthalmus japonicus* (Ocypodoidea), inhabits estuarine intertidal mud flats throughout the Indo-Pacific regions of Japan and Korea [24–27]. *M. japonicus* is also a major bioturbator of tidal flats and plays a crucial role in purifying sediment [28]. The movement of *M. japonicus* may affect the distribution of macro-infauna because of the accompanying disturbance in the physical and chemical properties of the sediment [25]. Previous studies have suggested that *M. japonicus* may be a useful indicator for modeling the exposure effects of chemical toxicity in the sediment environment [26,27,29,30].

In this study, we assessed the effects of crustacean embryogenesis and endocrine process by environmental chemical pollutants such as BPA, DEHP, and irgarol using the intertidal mud crab. To accomplish this, we characterized the *M. japonicus* VTG gene and analyzed phylogenetic relationships. The transcriptional responses of VTG and the tissue distribution of VTG mRNA expression were investigated from *M. japonicus* ovaries and hepatopancreases, plus other tissues, following exposure to either BPA, DEHP, or irgarol.

2. Materials and Methods

2.1. Organisms

M. japonicus was prepared from marine product markets in Yeosu (Jeonnam, Korea). On average, crabs ranged from 2.5 to 3.5 cm in carapace height, 2.7 to 4.3 cm in carapace width and 4 to 11 g in body weight. The crabs were placed in glass containers (45.7 × 35.6 × 30.5 cm) with natural seawater and aeration. All samples acclimatized for 1 d at 18 ± 1 °C in temperature, 25% salinity, and a 12 h light-dark schedule. Non-damaged crabs were selected. Experiments were performed based on the guidelines of the Chonnam National University Institutional Animal Care and Use Committee.

2.2. Exposure Experiments

Irgarol (2-(tert-butylamino)-4-(cyclo-propylamino)-6-(methyl-thio)-s-triazine) and BPA (99.9% pure) were prepared from Sigma-Aldrich (St. Louis, MO, USA). DEHP solutions were obtained from a solid compound (99%, Junsei Chemical Co. Ltd., Japan). The 10 mg L^{-1} stock solutions of BPA, DEHP, and irgarol were prepared by dissolving the chemicals in 99% acetone at room temperature. For working solutions of 1, 10, and 30 μg L^{-1} for each chemical; the stock solution was diluted with seawater. In solvent controls, a solvent concentration was <0.5% acetone.

For each chemical exposure, the crabs (*n* = 120) were randomly divided into four experimental groups (1, 10, and 30 μg L^{-1} treatment solutions, as well as solvent controls). Ten crabs were exposed with one of three doses of BPA, DEHP, or irgarol. Treatment times were 24 h, 96 h, and 7 d. Three individuals were subjected to tissue extractions for each time interval in each chemical treatment condition and control crab. The seawater was changed every day by adding the equivalent concentration of each chemical during the experiment. No food was provided during the experimental period. All experiments were run in three replicates with independent samples. Following exposure, tissue samples were immediately extracted from the crabs.

2.3. Characterization of the M. Japonicus Vitellogenin (VTG) Genes and Phylogenetic Analysis

VTG gene sequences were obtained using the 454 GS FLX transcriptomic database of the *M. japonicus* body [31]. The identified VTG cDNA sequences were compared with the VTG sequences in crustacean species available on the national center for biotechnology information (NCBI) database, using basic local alignment search tool (BLAST) searching (http://www.ncbi.nlm.nih.gov). The deduced amino acid sequences were obtained using the ExPASy translation program and aligned using the Clustal W2 tool.

To analyze the *M. japonicus* VTG gene in a phylogenetic tree, 14 crustacean VTG sequences were downloaded from NCBI for comparison of their similarities with the deduced amino acid sequences of the *M. japonicus* VTG. The GeneDoc Program (ver. 2.6.001) was used to display the multiple aligned sequences. A construction of the phylogenetic tree was made by neighbor-joining analyses of the Mega X program (version 10.04). The bootstrap value was calculated by 1000 replicates.

2.4. RNA Isolation and mRNA Expression Analysis

Total RNA from *M. japonicus* was obtained using the RNAiso Plus reagent (Takara, Dalian, China) following the manufacturer's protocol. Recombinant DNase I (RNase free) (Takara, Kusatsu, Japan) treatment was used to eliminate contamination of genomic DNA. Integrity and quantity of the extracted RNA were checked using 0.8% agarose gel for electrophoresis and a Nano-Drop 1000 (Thermo Fisher Scientific, Foster City, CA, USA). cDNA synthesis was performed with 3 μg of total RNA using the SuperScript™III RT kit (Invitrogen, Foster City, CA, USA) following the manufacturer's protocol. After synthesis, the diluted cDNA (50-fold) was stored at $-80\,°C$.

To evaluate VTG gene responses in multiple tissues, the total RNA was extracted from *M. japonicus* in a seawater control. Quantitative real-time PCR (polymerase chain reaction, RT-PCR) was performed using the Exicycler™96 (Bioneer, Daejeon, Korea) with the master mix (Bioneer, Daejeon, Korea). The glyceraldehyde-3-phosphate dehydrogenase (*Mj GAPDH*) gene of *M. japonicus* was used as an internal reference [16]. The primer sequences were: *Mj_*VTG forward 5′-CTTGGGCTCTCCAGTTC TTG-3′; *Mj_*VTG reverse 5′-CCACGTATGCCTCTTTTGGT-3′; *Mj_*GAPDH forward 5′-TGCTGATGCACCCATGTTTG-3′; *Mj_*GAPDH reverse 5′-AGGCCCTGGACAATCTCAA AG-3′. The size of the PCR product was 164 bp for the VTG gene and 147 bp for the GAPDH gene. The reaction mixture (20 μL) contained 6 μL of diluted cDNA, 10 μL of 2x SYBR green dye (Bioneer, Daejeon, Korea), 0.5 μL of forward and reverse primers (10 μM), and 3.0 μL of RNase-free water. The following RT-PCR cycle was used for amplification of the VTG gene: $94\,°C$ for 40 s, followed by 38 cycles of $94\,°C$ for 15 s, $53\,°C$ for 30 s and $72\,°C$ for 40 s. The Exicycler™96 real time system program (version 3.54.8) was used for the verification of the RT-PCR baseline. The calculation for relative transcript levels was used by the $2^{-\Delta\,\Delta ct}$ method [32] and normalized with GAPDH.

2.5. Statistical Analysis

Statistical analysis was conducted using statistical package for the social sciences (SPSS) 12.0 KO (SPSS Inc., Chicago, IL, USA). Statistical analysis for tissue distribution was performed using Dunnett's multiple range test of One-Way Analysis of Variance (ANOVA). An independent sample *t*-test was used to compare significant differences in the transcriptional expression of VTG between the ovaries and hepatopancreas under different chemical concentrations. A two-way ANOVA was conducted to assess the effects of dose and exposure to each chemical on VTG expression. Differences were statistically significant at $p < 0.05$ (*) and $p < 0.01$ (**). Data are presented as the mean $\pm$ SD.

3. Results

3.1. Characterization of the M. Japonicus VTG Gene

The partial cDNA of the *M. japonicus* VTG gene was obtained from the GS-FLX transcriptome database of the *M. japonicus* crab [31]. The *M. japonicus* VTG DNA was 3700 bp long, included an open reading frame (ORF) of 1202 amino acids, and included a lipoprotein amino acid terminal region of VTG (Figure 1). The *M. japonicus* VTG nucleotide sequence was 87% homologous with that of *Eriocheir sinensis* (KC699915). The predicted amino acid sequences of the VTG were 84%, 72%, 65%, and 63% homologous with those of *E. sinensis* (AGM75775), *Longpotamon honanense* (freshwater crab; AKI23633), *Scylla paramamosain* (green mud crab; ACO36035), and *Portunus trituberculatus* (swimming crab; AAX94762), respectively. Thus, the VTG gene of *M. japonicus* does not show a high homology with that of the other crabs (Figure 1). A phylogenetic analysis revealed two clades of VTG of the crabs (Figure 2). One clade was composed of VTGs from *M. japonicus*, *E. sinensis* and *L. honanense*. Another clade had homologous VTGs from *S. paramamosain*, *P. trituberculatus*, *Charybdis feriata* (Indo-Pacific crab; AAU93694), and *Callinectes sapidus* (blue crab; ABC41925). In addition, the black tiger shrimp (*Penaeus monodon*) (ABB89953) VTG formed another clade with the VTG of lobster (*Homarus americanus*; ABO09863), multiple prawn species, and other related shrimp species (Figure 2).

Figure 1. Vitellogenin (VTG) gene of the intertidal mud crab, *M. japonicus*. Multiple sequence alignments of the deduced *M. japonicus* VTG gene sequences with the homologous sequences of other crabs.

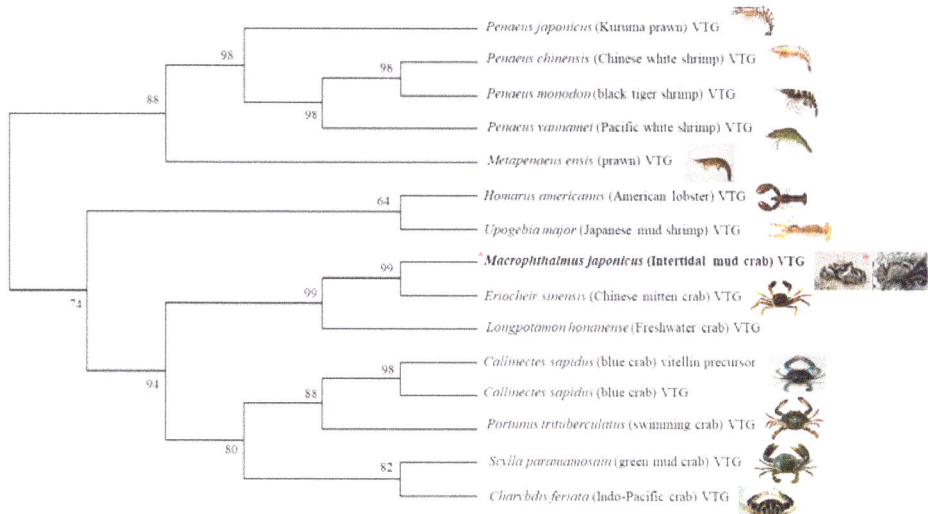

Figure 2. Phylogenetic tree of *M. japonicus* VTG gene constructed by neighbor-joining analysis (bootstrap value 1000). The numbers at the nodes are the percentage bootstrap values. GenBank accession numbers for VTGs are: *M. japonicus* VTG (Acc **), *Eriocheir sinensis* (KC699915), *Longpotamon honanense* (AKI23633), *Callinectes sapidus* vitellin precursor (AEI59132), *C. sapidus* VTG (ABC41925), *Portunus trituberculatus* (AAX94762), *Scylla paramamosain* (ACO36035), *Charybdis feriata* (AAU93694), *Homarus americanus* (ABO09863), *Upogebia major* (BAF91417), *Penaeus japonicus* (BAD98732), *Penaeus chinensis* (ABC86571), *Penaeus monodon* (ABB89953), *Penaeus vannamei* (AAP76571), and *Metapenaeus ensis* (AAN40700). ** *Accession #s to be deposited in GenBank.*

3.2. Basal Levels of M. Japonicus VTG Expression in Multiple Tissues

The basal expression levels of VTG were observed in six tissues (gill, hepatopancreas, muscle, ovaries, heart, and stomach) (Figure 3). The fold change of each VTG mRNA expression in each tissue was based on the VTG expression level in gill tissue (when the set value of the gill = 1). The *M. japonicus* VTG gene expression was expressed differentially in all investigated tissues. The high level of *M. japonicus* VTG expression was evident in the ovaries and hepatopancreas, whereas relatively low levels were observed in the gill, muscle, stomach, and heart. The *M. japonicus* VTG mRNA expression varied significantly between the crab gill and ovaries, or hepatopancreas ($p < 0.05$).

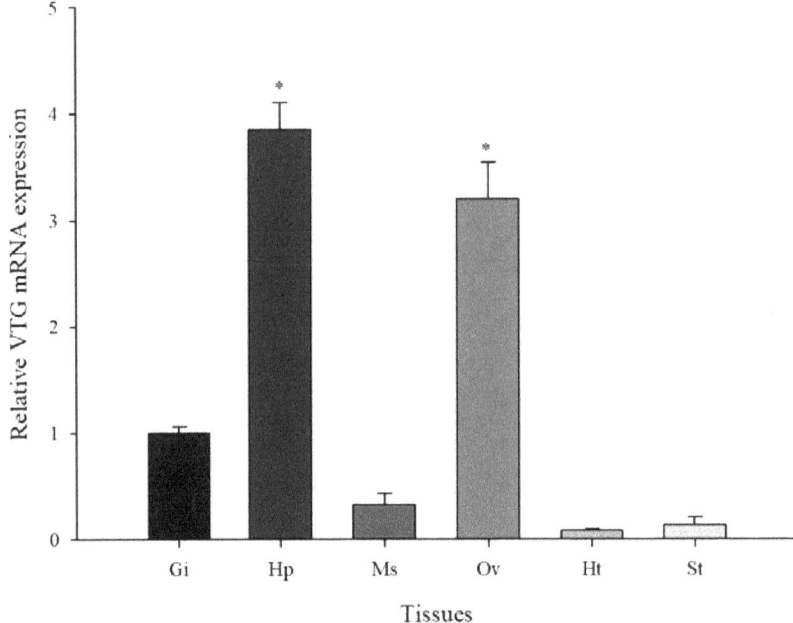

Figure 3. Basal expression of *M. japonicus* VTG transcripts among different tissues (Gi, gill; Hp, hepatopancreas; Ms, muscle; Ov, ovaries; Ht, heart; St, stomach). Each tissue was collected from 10 mud crabs. The experiments were performed in triplicate. The data are represented as the mean ± SD. The relative expression levels of each gene in each tissue were compared with the respective gene expression levels in gills (relative to set value of gill = 1). Significant differences at * $p < 0.05$ are indicated with an asterisk.

3.3. VTG Gene Expressions in M. Japonicus Ovaries and Hepatopancreas after DEHP, BPA, or Irgarol Exposures

After an exposure of 24 h, VTG mRNA expression generally increased in *M. japonicus* hepatopancreas exposed to DEHP, BPA or irgarol (Figure 4A). In particular, exposure to 1 and 30 µg L^{-1} DEHP significantly induced VTG gene expression ($p < 0.05$) relative to the control (when the value of the control = 1). In the ovaries, transcriptional levels of VTG increased slightly after BPA exposure (Figure 4B). In contrast, irgarol slightly decreased VTG expression at a relatively low concentration of 1 µg L^{-1}, although VTG response was somewhat induced at a relatively high irgarol concentration of 30 µg L^{-1}. After exposure to different concentrations of DEHP, VTG gene expression was significantly upregulated at 10 (2.8-fold) and 30 µg L^{-1} (3.3-fold) of DEHP in a concentration-dependent manner ($p < 0.05$).

The VTG gene response increased in hepatopancreas exposed to DEHP and BPA for 96 h, whereas a decrease in VTG expression resulted from irgarol exposure (Figure 5A). The up-regulation of VTG gene expression was significantly different at a relatively high concentration of BPA (30 µg L^{-1}) ($p < 0.05$). In addition, DEHP exposure significantly induced the up-regulation of the VTG expression in the hepatopancreas. The highest VTG expression was observed at 30 µg L^{-1} (4.9-fold) DEHP ($p < 0.01$). In contrast, after 96 h of irgarol exposure, VTG decreased in a dose-dependent manner. In the *M. japonicus* ovaries, BPA and DEHP exposure significantly upregulated transcriptional levels of VTG following exposure to all concentrations ($p < 0.05$) (Figure 5B). Similar to the VTG gene response in the hepatopancreas, the VTG expression in the *M. japonicus* ovaries downregulated after irgarol exposure for 96 h. The highest level of VTG expression was found at 30 µg L^{-1} (5.1-fold) of DEHP ($p < 0.01$).

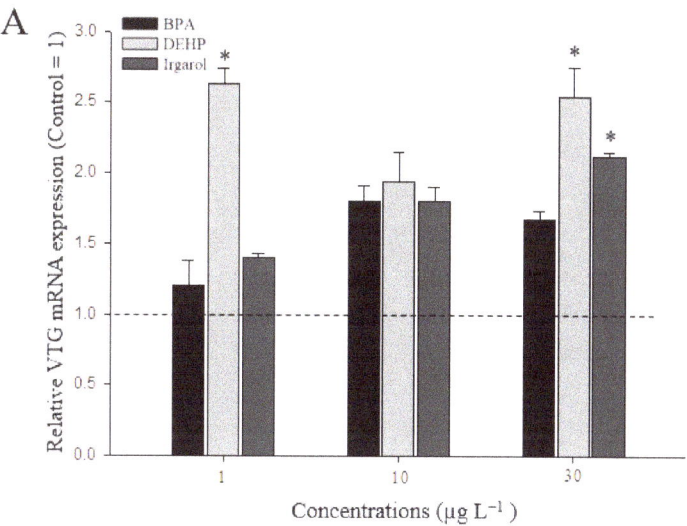

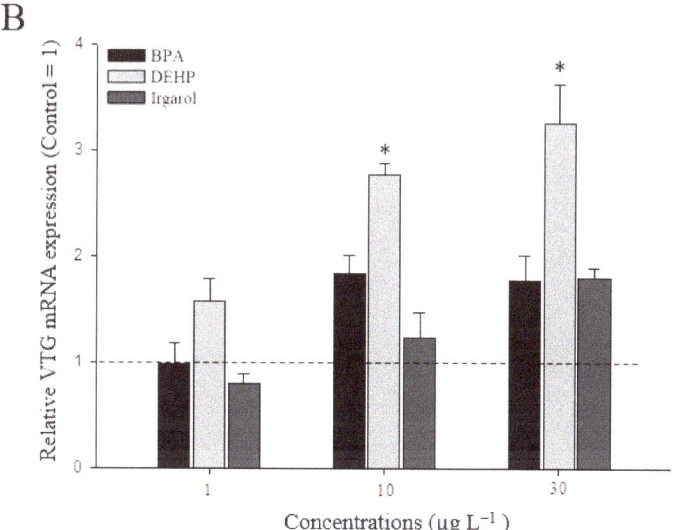

Figure 4. Transcriptional expression of VTG gene in *M. japonicus* hepatopancreas (**A**) and ovaries (**B**) exposed to each dose of bisphenol A (BPA), di(2-ethylhexyl) phthalate (DEHP), and irgarol for 24 h. The values were normalized against glyceraldehyde-3-phosphate dehydrogenase (GAPDH). Values of each bar represent the mean ± SD. A statistically significant difference is presented by an asterisk at * $p < 0.05$ as compare to the control (relative control value of VTG = 1).

At 7 d of exposure, there were different responses of the VTG gene between the hepatopancreas and ovary tissues (Figure 6). In the hepatopancreas, significant up-regulation of the VTG gene was observed at relatively low concentrations of 1 μg L^{-1} of BPA (4.3 fold) and DEHP (5.1 fold) (Figure 6A). The responses of VTG gene expression decreased in a concentration-dependent manner. After irgarol exposure, VTG mRNA expression slightly increased in the *M. japonicus* hepatopancreas, compared to the control level, for 7 d. In terms of VTG response in the ovaries, VTG expression was upregulated at all concentrations of BPA, DEHP, or irgarol exposure (Figure 6B). BPA exposure significantly increased VTG gene expression at 1 (4.3-fold), 10 (4.0-fold), and 30 μg L^{-1} (5.1-fold) of BPA ($p < 0.05$). DEHP treatment also significantly induced the VTG mRNA level at 1 (4.2 fold), 10 (5.3 fold) and 30 μg L^{-1} (7.2 fold) of DEHP ($p < 0.01$). The trend of VTG in response to DEHP exposure increased in a dose-dependent manner. In addition, significant expression of VTG was found at a relatively high concentration of 30 μg L^{-1} (3.8 fold) of irgarol, although VTG gene expression generally increased in *M. japonicus* ovaries exposed to irgarol for 7 d. Furthermore, there are no significant difference in weight changes between the control and treated groups after 7 d exposures to each chemical (Supplementary Figure S1).

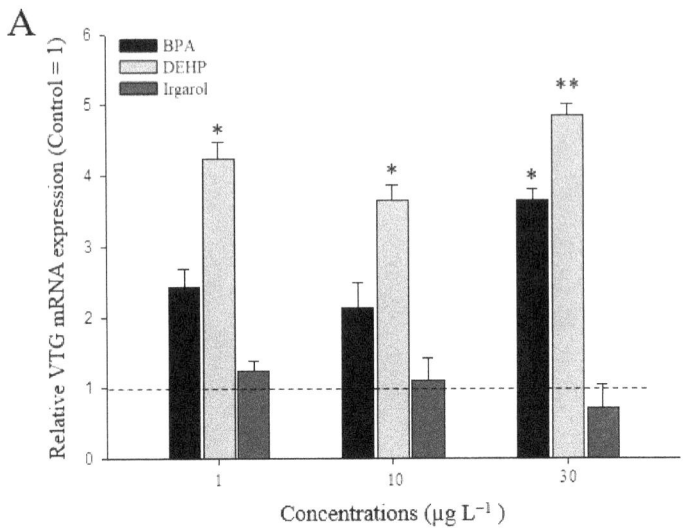

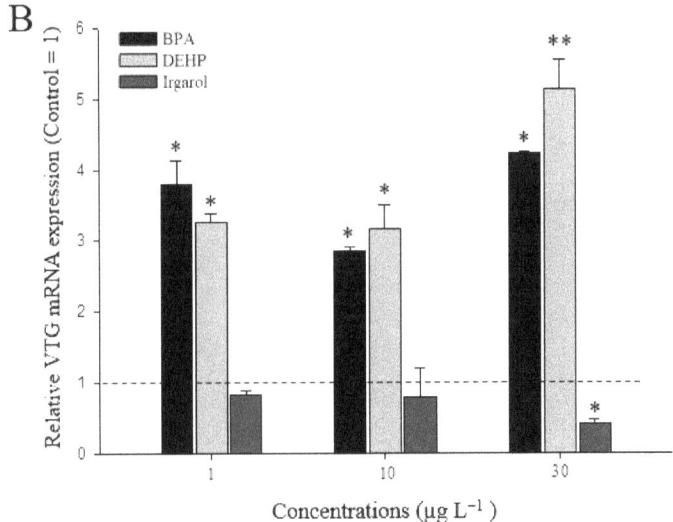

Figure 5. Expression of VTG mRNA in *M. japonicus* hepatopancreas (**A**) and ovaries (**B**) after exposure to each dose of BPA, DEHP, and irgarol for 96 h. The values were normalized against GAPDH. Values of each bar represent the mean ± SD. A statistically significant difference is presented by an asterisk at * $p < 0.05$ and ** $p < 0.01$ as compare to the control (relative control value of VTG = 1).

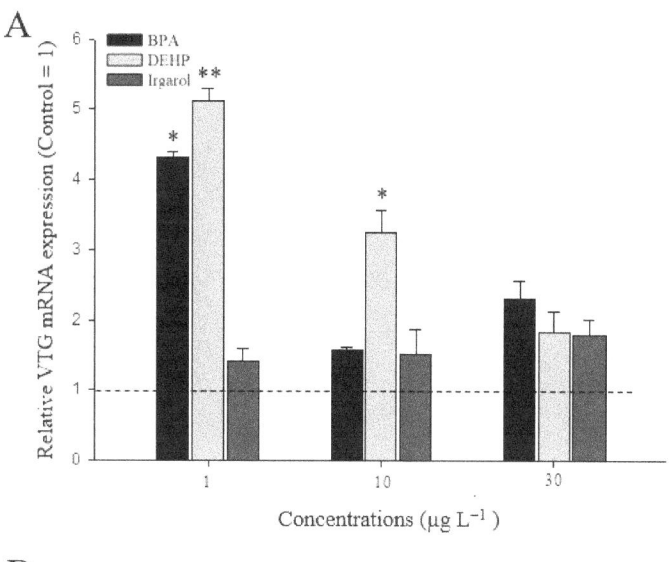

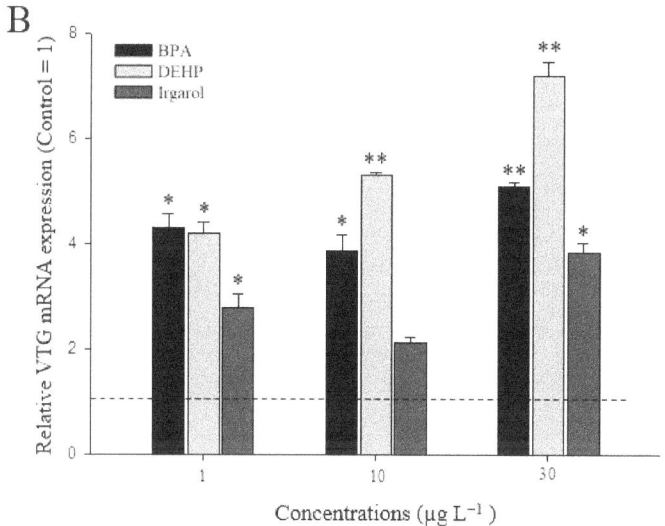

Figure 6. Expression of VTG mRNA in *M. japonicus* hepatopancreas (**A**) and ovaries (**B**) after exposure to each dose of BPA, DEHP, and irgarol for 7 d. The values were normalized against GAPDH. Values of each bar represent the mean ± SD. A statistically significant difference was presented by an asterisk at * $p < 0.05$ and ** $p < 0.01$ as compare to the control (relative control value of VTG = 1).

With different exposure times, VTG gene expression for 24 or 96 h did not differ significantly between the hepatopancreas and ovary tissues ($p > 0.05$). VTG expression levels for 7 d were significantly different between the hepatopancreas and gill ($p < 0.001$). An increase of the VTG gene expression was observed to be time-dependent in *M. japonicus* hepatopancreas exposed to the relatively low concentrations of 1 µg L^{-1} BPA and DEHP (Supplementary Figure S2). In ovary tissue, the up-regulations of VTG gene were found to be time-dependent at all concentrations of BPA and DEHP (Supplementary Figure S3).

4. Discussion

To understand VTG gene responses following exposure to environmental chemical pollutants, the cDNA sequence of *M. japonicus* VTG was identified and analyzed using phylogenetic analysis. The three functional domains of the crustacean VTG gene are the lipoprotein domain at the N-terminus (LPD_N), the DUF1943 domain, and the von Willebrand factors type-D domain (VWD) [2,23,33]. In this study, we characterized the *M. japonicus* VTG with 3700 bp ORF, which encodes a protein of 1202 amino acids. The deduced protein region of the *M. japonicus* VTG was predicted as the LPD_N of VTG, which is a conserved region found in various lipid transport proteins [34]. In multiple alignments of the amino acid sequences, *M. japonicus* VTG showed high homology with the protein from *E. sinensis*. The phylogenetic analysis revealed two clusters representing shrimps or crabs among the crustaceans. In these clusters, three clades were represented by crabs, including *M. japonicus*, other crabs, and *H. americanus*.

In this study, it was observed that *M. japonicus* mainly synthesizes VTG in the hepatopancreas and ovaries. In crustaceans, the hepatopancreas is a critical storage system of nutrients and energy in the forms of carbohydrates, proteins, and lipids, which are essential in regulating an energy balance during the reproductive cycle [35]. The hepatopancreas was also reported to be the main source of VTG activity in decapod species, although VTG expression was also found in the ovaries [5]. The primary expression site of the VTG gene in crustaceans has been a controversial topic. VTG production was only observed in the hepatopancreas of *Macrobrachium rosenbergii* (freshwater prawn), *Cherax quadricarinatus* (red claw crayfish), *Pandalus hypsinotus* (coonstriped shrimp) and *P. trituberculatus* [2,36–38]. In contrast, the VTG gene was expressed in both the hepatopancreas and ovaries of *Upogebia major* (mud shrimp) and marine crabs such as *S. paramamosain*, *Carcinus maenas* (green crab), *C. sapidus*, *C. japonica* and *E. sinensis* [9,23,39–41]. We identified that the major sites for VTG synthesis in *M. japonicus* are the hepatopancreas and ovaries.

The endocrine disruptor BPA is presently considered a ubiquitous emerging pollutant in aquatic environments. In this study, exposures to BPA triggered up-regulation of *M. japonicus* VTG expression in the hepatopancreas and ovary tissues for different exposure periods. BPA exposure increased VTG levels in various fish, such as fathead minnows, goldfish, and bluegills [42]. BPA toxicity at lower concentrations (<15 µg L^{-1}) showed induced ovarian expression of a steroidogenic gene involving VTG in *G. rarus* [13]. In addition, different modulations of VTG were reported in two VTG genes in the gilthead bream *Sparus aurata* and *G. rarus* [43,44]. These results indicate that induction of VTG following BPA exposure is a major response to the stressful effects of endocrine-disruption.

The endocrine effects of DEHP, a distinctive EDC, are noticeable in marine ecosystems where high concentrations of DEHP have been found in river estuaries and coastal areas [45]. In our study, DEHP exposure generally increased *M. japonicus* VTG expression in the tested tissues. The up-regulation of VTG gene expression in *M. japonicus* was more sensitive to DEHP exposure than to exposure to the other chemicals tested. In addition, transcriptional responses of the VTG gene in the hepatopancreas were found to be significant under relatively low concentrations of DEHP for the different exposure times, whereas VTG was expressed in the ovaries under all concentrations of DEHP tested. Adeogun reported that DEHP exposure induced time- and concentration-dependent increases of VTG levels in *C. gariepinus* [17]. Long exposure to low doses of DEHP reduced oocyte maturation [46] and increased the levels of hepatic VTG transcripts in *G. rarus* [16]. In addition, VTG levels in marine medaka livers increased after DEHP exposure [47]. These results strongly emphasize the suitability of using the VTG gene as a biomarker to assess the environmental presence of DEHP.

Since the use of irgarol as an algaecide has increased, irgarol has been detected frequently in coastal environments [48]. However, there is not enough information about the potential effects of irgarol exposure on the endocrine system of crustaceans. In our study, the responses of VTG gene expression following exposure to irgarol were different. After a 24 h exposure to irgarol, the *M. japonicus* VTG gene was expressed in a dose-dependent manner. The up-regulation of VTG relative to the control decreased in the hepatopancreas and ovaries when exposed to irgarol for 96 h. VTG gene expression recovered after 7 d of irgarol exposure. These results are the first report of the endocrine effects of irgarol exposure in crustaceans. A recent study reported that irgarol exposure induced changes in thyroid endpoints in the inland silverside fish (*Menidia beryllina*) under salinity and temperature changes [28]. In *A. tenuis*, exposure to irgarol for 7 d resulted in body color changes and reduced transcriptional expression of the HSP90 gene [29]. In a previous study, irgarol exposure induced expression changes of digestion related genes and chitinase genes [17]. Thus, VTG gene expression was affected by environmental chemical pollutants such as BPA, DEHP, and irgarol. After EDC exposures, the expression pattern of the *M. japonicus* VTG gene was correlated with short-term EDC exposure in the hepatopancreas ($p = 0.791$) and long-term EDC exposure in ovary tissue ($p = 0.927$).

In conclusion, our results demonstrate that *M. japonicus* is sensitive to several chemical pollutants, as measured by changes of the VTG gene expression in the hepatopancreas and ovaries. The results suggest that the molecular balance of VTG expression levels of intertidal mud crabs appear to be affected by EDCs such as BPA, DEHP, and irgarol at even relatively low concentrations of $1 \mu g \ L^{-1}$. The expression levels of the VTG gene are important factors to be considered for evaluating xenoestrogenic responses. The change of the VTG responses are related with disturbed processes of reproductive physiology such as egg yolk generation in *M. japonicus*, because of the up-regulation of VTG expression in the ovaries after EDC exposures. The crab hepatopancreas is an organ for VTG synthesis and storage of energy accumulation, which can support lipid transportation for oogenesis [49]. EDC exposure also impaired the molecular balance of the VTG level through the disturbance of energy metabolisms in *M. japonicus* hepatopancreas. Up or down-regulation of VTG gene may result in sensitive defense activities of *M. japonicus* crabs to potential EDCs pollutants, given the major function of VTG in stabilizing the endocrine process and triggering lipid uptake for energy storage during crustacean embryogenesis. Our findings suggest that *M. japonicus* VTG may be a useful key to estimate the endocrine effects of antifoulants such as irgarol, BPA, and DEHP. In further studies, we will need to demonstrate the effects of one or more combined exposures of these EDCs in VTG protein synthesis.

Supplementary Materials: The following are available online at http://www.mdpi.com/2076-3417/9/7/1401/s1, Figure S1: The relative weight (g) in *M. japonicus* exposed to each dose of BPA, DEHP, and irgarol for 7 d. Figure S2: Transcriptional expression of VTG gene in *M. japonicus* hepatopancreas exposed to each dose of BPA, DEHP, and irgarol for each exposure period. Figure S3: Transcriptional expression of VTG gene in *M. japonicus* ovary exposed to each dose of BPA, DEHP, and irgarol for each exposure period.

Author Contributions: Conceptualization: K.P. and I.-S.K.; Methodology: K.P and I.-S.K.; Formal Analysis: K.P., H.J. and D.-K.K.; Investigation: K.P., D.-K.K. and H.J.; Resources: K.P. and I.-S.K.; Writing—Original Draft Preparation: K.P.; Writing—Review and Editing: H.J., D.-K.K. and I.-S.K.; Supervision: I.-S.K.; Project Administration: I.-S.K.; Funding Acquisition: I.-S.K.

Funding: This study was supported by the National Research Foundation of Korea, Korea, which is funded by the Korean Government (NRF-2018-R1A6A1A-03024314).

Conflicts of Interest: The authors declare no conflict of interest.

References

1. Tsukimura, B.; Waddy, S.L.; Vogel, J.M.; Linder, C.J.; Bauer, D.K.; Borst, D.W. Characterization and quantification of yolk proteins in the lobster, *Homarus americanus*. *J. Exp. Zool.* **2002**, *292*, 367–375. [CrossRef] [PubMed]
2. Yang, F.; Xu, H.T.; Dai, Z.M.; Yang, W.J. Molecular characterization and expression analysis of vitellogenin in the marine crab *Portunus trituberculatus*. *Comp. Biochem. Physiol. B Biochem. Mol. Biol.* **2005**, *142*, 456–464. [CrossRef]
3. Okumura, T.; Yamano, K.; Sakiyama, K. Vitellogenin gene expression and hemolymph vitellogenin during vitellogenesis, final maturation, and oviposition in female kuruma prawn, *Marsupenaeus japonicus*. *Comp. Biochem. Physiol. A Mol. Integr. Physiol.* **2007**, *147*, 1028–1037. [CrossRef] [PubMed]
4. Subramoniam, T. Mechanisms and control of vitellogenesis in crustaceans. *Fish. Sci.* **2011**, *77*, 1–21. [CrossRef]
5. Jia, X.; Chen, Y.; Zou, Z.; Lin, P.; Wang, Y.; Zhang, Z. Characterization and expression profile of Vitellogenin gene from *Scylla paramamosain*. *Gene* **2013**, *520*, 119–130. [CrossRef]

6.   Thongda, W.; Chung, J.S.; Tsutsui, N.; Zmora, N.; Katenta, A. Seasonal variations in reproductive activity of the blue crab, *Callinectes sapidus*: Vitellogenin expression and levels of vitellogenin in the hemolymph during ovarian development. *Comp. Biochem. Physiol. A Mol. Integr. Physiol.* **2015**, *179*, 35–43. [CrossRef] [PubMed]

7.   Matozzo, V.; Gagne, F.; Marin, M.G.; Ricciardi, F.; Blaise, C. Vitellogenin as a biomarker of exposure to estrogenic compounds in aquatic invertebrates: A review. *Environ. Int.* **2008**, *34*, 531–545. [CrossRef] [PubMed]

8.   Park, K.; Kwak, T.S.; Kwak, I.S. Vitellogenin Gene Characterization and Expression of Asian Paddle Crabs (*Charybdis japonica*) following Endocrine Disrupting Chemicals. *Ocean Sci. J.* **2014**, *49*, 127–135. [CrossRef]

9.   Jeon, J.M.; Lee, S.O.; Kim, K.S.; Baek, H.J.; Kim, S.; Kim, I.K.; Mykles, D.L.; Kim, H.W. Characterization of two vitellogenin cDNAs from a Pandalus shrimp (*Pandalopsis japonica*): Expression in hepatopancreas is down-regulated by endosulfan exposure. *Comp. Biochem. Physiol. B Biochem. Mol. Biol.* **2010**, *157*, 102–112. [CrossRef] [PubMed]

10. Neris, J.B.; Olivares, D.M.M.; Velasco, F.G.; Luzardo, F.H.M.; Correia, L.O.; González, L.N. HHRISK: A code for assessment of human health risk due to environmental chemical pollution. *Ecotoxicol. Environ. Saf.* **2018**, *170*, 538–547. [CrossRef] [PubMed]

11. Shafei, A.; Ramzy, M.M.; Hegazy, A.I.; Husseny, A.K.; El-Hadary, U.G.; Taha, M.M.; Mosa, A.A. The molecular mechanisms of action of the endocrine disrupting chemical bisphenol A in the development of cancer. *Gene* **2018**, *647*, 235–243. [CrossRef]

12. Kolpin, D.W.; Furlong, E.T.; Meyer, M.T.E.; Michael, T.; Zaugg, S.D.; Barber, L.B.; Buxton, H.T. Pharmaceuticals, hormones, and other organic wastewater contaminants in U.S.streams, 1999–2000: A national reconnaissance. *Environ. Sci. Technol.* **2002**, *36*, 1202–1211. [CrossRef] [PubMed]

13. Dong, J.; Xiang-Li, L.I.; Liang, R.J. Bisphenol A pollution of surface water and its environmental factors. *J. Ecol. Rural Environ.* **2009**, *25*, 94–97.

14. Park, K.; Kwak, I.S. Characterize and Gene Expression of Heat Shock Protein 90 in Marine Crab *Charybdis japonica* following Bisphenol A and 4-Nonylphenol Exposures. *Environ. Health Toxicol.* **2014**, *29*, e2014002. [CrossRef] [PubMed]

15. Baek, J.H.; Gu, M.B.; Sang, B.I.; Kwack, S.J.; Kim, K.B.; Lee, B.M. Risk reduction of adverse effects due to di-(2-ethylhexyl) phthalate (DEHP) by utilizing microbial degradation. *J. Toxicol. Environ. Health. A* **2009**, *72*, 1388–1394. [CrossRef] [PubMed]

16. Wang, X.; Yang, Y.; Zhang, L.; Ma, Y.; Han, J.; Yang, L.; Zhou, B. Endocrine disruption by di-(2-ethylhexyl)-phthalate in Chinese rare minnow (*Gobiocypris rarus*). *Environ. Toxicol. Chem.* **2013**, *32*, 1846–1854. [CrossRef]

17. Adeogun, A.O.; Ibor, O.R.; Imiuwa, M.E.; Omogbemi, E.D.; Chukwuka, A.V.; Omiwole, R.A.; Arukwe, A. Endocrine disruptor responses in African sharptooth catfish (*Clarias gariepinus*) exposed to di-(2-ethylhexyl)-phthalate. *Comp. Biochem. Physiol. C Toxicol. Pharmacol.* **2018**, *213*, 7–18. [CrossRef]

18. Li, X.; Yin, P.; Zhao, L. Phthalate esters in water and surface sediments of the Pearl River Estuary: Distribution, ecological, and human health risks. *Environ. Sci. Pollut. Res. Int.* **2016**, *23*, 19341–19349. [CrossRef]

19. Sapozhnikova, Y.; Wirth, E.; Schiff, K.; Fulton, M. Antifouling biocides in water and sediments from California marinas. *Mar. Pollut. Bull.* **2013**, *69*, 189–194. [CrossRef] [PubMed]

20. Moreira, L.B.; Diamante, G.; Giroux, M.; Xu, E.G.; Abessa, D.M.S.; Schlenk, D. Changes in thyroid status of *Menidia beryllina* exposed to the antifouling booster irgarol: Impacts of temperature and salinity. *Chemosphere* **2018**, *209*, 857–865. [CrossRef] [PubMed]

21. Ishibashi, H.; Minamide, S.; Takeuchi, I. Identification and characterization of heat shock protein 90 (HSP90) in the hard coral *Acropora tenuis* in response to Irgarol 1051. *Mar. Pollut. Bull.* **2018**, *133*, 773–780. [CrossRef] [PubMed]

22. Konstantinou, I.K.; Albanis, T.A. Worldwide occurrence and effects of antifouling paint booster biocides in the aquatic environment: A review. *Environ. Int.* **2004**, *30*, 235–248. [CrossRef]

23. Li, L.; Li, X.J.; Wu, Y.M.; Yang, L.; Li, W.; Wang, Q. Vitellogenin regulates antimicrobial responses in Chinese mitten crab, *Eriocheir sinensis. Fish Shellfish Immunol.* **2017**, *69*, 6–14. [CrossRef] [PubMed]

24. Kitaura, J.; Nishida, M.; Wada, K. Genetic and behavioral diversity in the *Macrophthalmus japonicus* species complex (Crustacea: Brachyura: Ocypodidae). *Mar. Biol.* **2002**, *140*, 1–8.

25. Tanaka, Y.; Aoki, S.; Okamoto, K. Effects of the bioturbating crab *Macrophthalmus japonicus* on abiotic and biotic tidal mudflat characteristics in the Tama River, Tokyo Bay, Japan. *Plankton. Benthos. Res.* **2017**, *12*, 34–43. [CrossRef]

26. Nikapitiya, C.; Kim, W.S.; Park, K.; Kim, J.; Lee, M.O.; Kwak, I.S. Chitinase gene responses and tissue sensitivity in an intertidal mud crab (*Macrophthalmus japonicus*) following low or high salinity stress. *Cell Stress Chaperones* **2015**, *20*, 517–526. [CrossRef]

27. Park, K.; Kwak, T.S.; Kim, W.S.; Kwak, I.S. Changes in exoskeleton surface roughness and expression of chitinase genes in mud crab *Macrophthalmus japonicus* following heavy metal differences of estuary. *Mar. Pollut. Bull.* **2019**, *138*, 11–18. [CrossRef]

28. Park, K.; Nikapitiya, C.; Kwak, T.S.; Kwak, I.S. Antioxidative-related Genes Expression Following Perfluorooctane Sulfonate (PFOS) Exposure in the Intertidal Mud Crab, *Macrophthalmus japonicus*. *Ocean Sci. J.* **2015**, *50*, 547–556. [CrossRef]

29. Nikapitiya, C.; Kim, W.S.; Park, K.; Kwak, I.S. Identification of potential markers and sensitive tissues for low or high salinity stress in an intertidal mud crab (*Macrophthalmus japonicus*). *Fish Shellfish Immunol.* **2014**, *41*, 407–416. [CrossRef]

30. Park, K.; Nikapitiya, C.; Kim, W.S.; Kwak, T.S.; Kwak, I.S. Changes of exoskeleton surface roughness and expression of crucial participation genes for chitin formation and digestion in the mud crab (*Macrophthalmus japonicus*) following the antifouling biocide irgarol. *Ecotoxicol. Environ. Saf.* **2016**, *132*, 186–195. [CrossRef] [PubMed]

31. Park, K.; Nikapitiya, C.; Kwak, I.S. Identification and expression of proteolysis response genes for *Macrophthalmus japonicus* exposure to irgarol toxicity. *Ann. Limnol. Int. J. Limnol.* **2016**, *52*, 65–74. [CrossRef]

32. Livak, K.J.; Schmittgen, T.D. Analysis of relative gene expression data using real time quantitative PCR and the 2_ΔΔCt method. *Methods* **2001**, *25*, 402–408. [CrossRef]

33. Tiu, S.H.; Hui, H.L.; Tsukimura, B.; Tobe, S.S.; He, J.G.; Chan, S.M. Cloning and expression study of the lobster (*Homarus americanus*) vitellogenin: Conservation in gene structure among decapods. *Gen. Comp. Endocrinol.* **2009**, *160*, 36–46. [CrossRef]

34. Anderson, T.A.; Levitt, D.G.; Banaszak, L.J. The structural basis of lipid interactions in lipovitellin, a soluble lipoprotein. *Structure* **1998**, *6*, 895–909. [CrossRef]

35. Yang, J.; Sun, H.; Qian, Y.; Yang, J. Impairments of cadmium on vitellogenin accumulation in the hepatopancreas of freshwater crab *Sinopotamon henanense*. *Environ. Sci. Pollut. Res. Int.* **2017**, *24*, 18160–18167. [CrossRef]

36. Okuno, A.; Yang, W.J.; Jayasankar, V.; Saido-Sakanaka, H.; Huong, D.T.; Jasmani, S.; Atmomarsono, M.; Subramoniam, T.; Tsutsui, N.; Ohira, T.; et al. Deduced primary structure of vitellogenin in the giant freshwater prawn, *Macrobrachium rosenbergii*, and yolk processing during ovarian maturation. *J. Exp. Zool.* **2002**, *292*, 417–429. [CrossRef] [PubMed]

37. Abdu, U.; Davis, C.; Khalaila, I.; Sagi, A. The vitellogenin cDNA of *Cherax quadricarinatus* encodes a lipoprotein with calcium binding ability, and its expression is induced following the removal of the androgenic gland in a sexually plastic system. *Gen. Comp. Endocrinol.* **2002**, *127*, 263–272. [CrossRef]

38. Tsutsui, N.; Saido-Sakanaka, H.; Yang, W.J.; Jayasankar, V.; Jasmani, S.; Okuno, A.; Ohira, T.; Okumura, T.; Aida, K.; Wilder, M.N. Molecular characterization of a cDNA encoding vitellogenin in the coonstriped shrimp, *Pandalus hypsinotus* and site of vitellogenin mRNA expression. *J. Exp. Zool. A Ecol. Genet. Physiol.* **2004**, *301*, 802–814. [CrossRef]

39. Zmora, N.; Trant, J.; Chan, S.M.; Chung, J.S. Vitellogenin and its messenger RNA during ovarian development in the female blue crab, *Callinectes sapidus*: Gene expression, synthesis, transport, and cleavage. *Biol. Reprod.* **2007**, *77*, 138–146. [CrossRef]

40. Kang, B.J.; Nanri, T.; Lee, J.M.; Saito, H.; Han, C.H.; Hatakeyama, M.; Saigusa, M. Vitellogenesis in both sexes of gonochoristic mud shrimp, *Upogebia major* (Crustacea): Analyses of vitellogenin gene expression and vitellogenin processing. *Comp. Biochem. Physiol. B Biochem. Mol. Biol.* **2008**, *149*, 589–598. [CrossRef]

41. Ding, X.; Nagaraju, G.P.C.; Novotney, D.; Lovett, D.L.; Borst, D.W. Yolk protein expression in the green crab, *Carcinus maenas*. *Aquaculture* **2010**, *298*, 325–331. [CrossRef]

42. Allner, B.; Hennies, M.; Lerche, C.F.; Schmidt, T.; Schneider, K.; Willner, M.; Stahlschmidt-Allner, P. Kinetic determination of vitellogenin induction in the epidermis of cyprinid and perciform fishes: Evaluation of sensitive enzyme-linked immunosorbent assays. *Environ. Toxicol. Chem.* **2016**, *35*, 2916–2930. [CrossRef]

43. Maradonna, F.; Nozzi, V.; Dalla Valle, L.; Traversi, I.; Gioacchini, G.; Benato, F.; Colletti, E.; Gallo, P.; Di Marco Pisciottano, I.; Mita, D.G.; et al. A developmental hepatotoxicity study of dietary bisphenol A in *Sparus aurata* juveniles. *Comp. Biochem. Physiol. C Toxicol. Pharmacol.* **2014**, *166*, 1–13. [CrossRef] [PubMed]

44. Zhang, Y.; Gao, J.; Xu, P.; Yuan, C.; Qin, F.; Liu, S.; Zheng, Y.; Yang, Y.; Wang, Z. Low-dose bisphenol A disrupts gonad development and steroidogenic genes expression in adult female rare minnow *Gobiocypris rarus*. *Chemosphere* **2014**, *112*, 435–442. [CrossRef]

45. Heindler, F.M.; Alajmi, F.; Huerlimann, R.; Zeng, C.; Newman, S.J.; Vamvounis, G.; van Herwerden, L. Toxic effects of polyethylene terephthalate microparticles and Di(2-ethylhexyl)phthalate on the calanoid copepod, *Parvocalanus crassirostris*. *Ecotoxicol. Environ. Saf.* **2017**, *141*, 298–305. [CrossRef] [PubMed]

46. Guo, Y.; Yang, Y.; Gao, Y.; Wang, X.; Zhou, B. The impact of long term exposure to phthalic acid esters on reproduction in Chinese rare minnow (*Gobiocypris rarus*). *Environ. Pollut.* **2015**, *203*, 130–136. [CrossRef]

47. Ye, T.; Kang, M.; Huang, Q.; Fang, C.; Chen, Y.; Shen, H.; Dong, S. Exposure to DEHP and MEHP from hatching to adulthood causes reproductive dysfunction and endocrine disruption in marine medaka (*Oryzias melastigma*). *Aquat. Toxicol.* **2014**, *146*, 115–126. [CrossRef] [PubMed]

48. Bao, V.W.W.; Leung, K.M.Y.; Lui, G.C.S.; Lam, M.H.W. Acute and chronic toxicities of Irgarol alone and in combination with copper to the marine copepod *Tigriopus japonicus*. *Chemosphere* **2013**, *90*, 1140–1148. [CrossRef]

49. Chen, X.; Wang, J.; Hou, X.; Yue, W.; Huang, S.; Wang, C. Tissue expression profiles unveil the gene interaction of hepatopancreas, eyestalk, and ovary in the orecocious female Chinese mitten crab, *Eriocheir sinensis*. *BMC Genet.* **2019**, *20*, 12. [CrossRef]

Article

Health Risk Assessment of Banned Veterinary Drugs and Quinolone Residues in Shrimp through Liquid Chromatography–Tandem Mass Spectrometry

Ming-Yang Tsai [1,2,†], Chuen-Fu Lin [3,†], Wei-Cheng Yang [4], Chien-Teng Lin [3], Kuo-Hsiang Hung [2] and Geng-Ruei Chang [3,*]

[1] Animal Industry Division, Livestock Research Institute, Council of Agriculture, Executive Yuan, Tainan 71246, Taiwan; mytsai@mail.tlri.gov.tw
[2] Graduate Institute of Bioresources, National Pingtung University of Science and Technology, Pingtung 91201, Taiwan; khhung424@mail.npust.edu.tw
[3] Department of Veterinary Medicine, National Chiayi University, Chiayi 60054, Taiwan; cflin@mail.ncyu.edu.tw (C.-F.L.); vet540423@gmail.com (C.-T.L.)
[4] Department of Veterinary Medicine, School of Veterinary Medicine, National Taiwan University, Taipei 10617, Taiwan; yangweicheng@ntu.edu.tw
* Correspondence: grchang@mail.ncyu.edu.tw; Tel.: +886-5-2732959; Fax: +886-5-2732917
† These authors contributed equally to this work.

Received: 24 May 2019; Accepted: 13 June 2019; Published: 17 June 2019

Abstract: The presence of antibiotic residues in seafood and their effect on public health constitute a matter of concern for consumers worldwide. Antibiotic residues can have adverse effects on both humans and animals, especially residues of banned veterinary drugs. In this study, we applied a validated method to analyze veterinary drug residues in shrimp, including the levels of banned chloramphenicol, malachite green, leucomalachite green, and four nitrofuran metabolites as well as thiamphenicol, florfenicol, and five quinolones, which have no recommended maximum residual levels in shrimp tissues in Taiwan. We collected 53 samples of whiteleg, grass, or giant river shrimp from Taiwanese aquafarms and production areas from July 2016 to December 2017. We found 0.31 ng/g of a chloramphenicol in one grass shrimp, 5.62 ng/g of enrofloxacin in one whiteleg shrimp, 1.52 ng/g of flumequine in one whiteleg shrimp, and 1.01 ng/g of flumequine in one giant river shrimp, indicating that 7.55% of the samples contained veterinary drug residues. We evaluated the health risk by deriving the estimated daily intake (EDI). The quinolone residue EDI was below 1.0% of the acceptable daily intake recommended by the United Nations Food and Agriculture Organization and World Health Organization. The risk was thus discovered to be negligible, indicating no immediate health risk associated with shrimp consumption. The present findings can serve as a reference regarding food safety and in monitoring of the veterinary drug residues present in aquatic organisms. Continual monitoring of residues in shrimp is critical for further assessment of possible effects on human health.

Keywords: veterinary drug; residue; shrimp; mass spectrometry; risk assessment

---

1. Introduction

Taiwan has a geographic location and environment conducive to aquaculture development. Aquaculture in Taiwan has a long history of more than three centuries, and it has rapidly expanded, diversified, intensified, and technologically advanced from 1960 to the 1990s [1]. Despite Taiwan's land and water resource limitations, it is one of the major aquaculture producers in the world; therefore, it was once called the "kingdom of aquaculture" [2]. Until now, over 35 major and candidate species have been cultured for commercial purposes [3]. The average revenue of Taiwan from aquaculture reached US $11 billion between 2010 and 2015 [3]. Moreover, the annual aquaculture production was approximately 300,000 t during the 2010s. Specifically, shrimp culture production has been notable because Taiwan's government has strongly supported aquaculture since the 1980s [4], in particular, aquaculture of shrimp—including whiteleg shrimp (*Litopenaeus vannamei*), grass shrimp (*Penaeus monodon*), giant river shrimp (*Macrobrachium rosenbergii*), sand shrimp (*Metapenaeus ensis*), and kuruma shrimp (*Penaeus japonicus*). Moreover, the average revenue from the shrimp aquaculture industry reached US $3000 million between 2010 and 2016.

Intensive and large-scale breeding is preferred by aquaculture farmers in Taiwan because the farmers are limited to land use [5]. Moreover, their farms are generally situated near residential and agricultural areas, which makes biological control difficult [6]. The cultured species became more susceptible to bacterial, viral, parasitical, and fungal infections, necessitating the use of various veterinary drugs for the prevention and treatment of these infections. However, when such drugs are heavily employed in aquaculture, aquaculture products may contain drug residues, potentially exposing the consumers of the products to these residues. This has an inadvertent ecological impact and raises health concerns such as increased risk of allergies, carcinogen exposure, and the development of bacteria resistant to antibiotics [5]. Because of their benthic feeding behavior, shrimp can be used to indicate the levels of chemicals in aquatic environments [7]. Analyzing the amounts of pollutants in shrimp indicates the environmental levels of veterinary drugs and the extent to which drugs are transferred through the trophic chain.

In Taiwan, shrimp are the crustaceans cultured most extensively in land-based ponds [3], and in the inner regions of the island, shrimp are generally bred in a mix with other aquatic products. Typically, veterinary drugs are used to prevent or treat diseases in nonshrimp targets, which pollutes water and soil environments while contaminating shrimp. In this regard, shrimp products have played a crucial role in seafood safety. Residues of veterinary drugs in shrimp are a crucial concern regarding public health, especially when the residues are of banned chemicals that have been employed illegally. Therefore, in this study, we detected in shrimp samples the levels of the following banned veterinary compounds in Taiwan: leucomalachite green (LMG), malachite green (MG), nitrofuran metabolites, and chloramphenicol. These compounds' maximum residue limits (MRLs) in shrimp have not been established by the Taiwan Food and Drug Administration (TFDA). In addition, we detected florfenicol, thiamphenicol, and quinolone residues, including danofloxacin, difloxacin, enrofloxacin, flumequine, and sarafloxacin, in shrimp. The TFDA defined the MRL levels of these quinolones in livestock and chicken in 2018; however, the use of these compounds in the cultivation of shrimp is banned. The present study's detection of the residues of these compounds in shrimp thus indicates the degree of legal compliance regarding the use of these products. In addition, the seafood-consumption-based estimated daily intake (EDI) of these contaminants was determined to identify the effects of exposure to these veterinary drugs on the health of the Taiwanese public. Results were applied for assessing the risk of exposure to veterinary drugs among consumers in Taiwan. The findings of this study are useful when conducting evaluations of seafood safety and may be used as a reference among health authorities for establishing regulations.

2. Materials and Methods

2.1. Samples

Shrimp samples were obtained between July 2016 and December 2017 from aquafarms in the principal production areas (Yunlin, Chiayi, Tainan, Kaohsiung, and Pingtung). In total, 53 samples (23 whiteleg, 16 grass, and 14 giant river shrimps) were collected. These shrimps are bred on a large scale in Taiwan [3]. We removed, homogenized, and stored the soft tissues of all shrimp samples at −20 °C until they were analyzed.

2.2. Chemicals and Reagents

Analytical compounds included veterinary drugs, namely chloramphenicol (98.6%), thiamphenicol (98.5%), florfenicol (98%), MG (98.0%), and LMG (99.0%), purchased from Dr. Ehrenstorfer GmbH (Ausburg, Germany). Nitrofuran metabolites, namely 5-methylmorpholino-3-amino-2-oxazolidinone (AMOZ, 99.9%), 3-amino-2-oxazolidinone (AOZ, 99.7%), and 1-aminohydantoin hydrochloride (99.9%), were obtained from Sigma-Aldrich (St. Louis, MO, USA). Semicarbazide hydrochloride (99.5%) was obtained from Chem Service Inc. (West Chester, PA, USA). Danofloxacin (98.0%), difloxacin (98.0%), enrofloxacin (98.0%), flumequine (98.0%), and sarafloxacin (95.0%) were obtained from Sigma-Aldrich. In addition, stable isotopically labeled internal standards, AMOZ-D5 (99.0%) and AOZ-D4 (99.0%), were purchased from Dr. Ehrenstorfer GmbH. Other internal standards, namely SC-13C15N2 (99%), MG-D5 picrate (99.9%), and LMG-D5 (99.8%), were purchased from Sigma-Aldrich.

Chromatography-grade acetonitrile (ACN), acetone, ammonium acetate, dipotassium phosphate, ethyl acetate (EtOAc), formic acid (FA), hydrochloric acid (HCl), methanol (MeOH), n-hexane, and sodium hydroxide (NaOH) were purchased from Merck (Darmstadt, Germany). Reagent-grade 2-nitrobenzaldehyde (2-NBA), sodium chloride (NaCl), and *N,N,N',N'*-tetramethyl-1,4-phenylenediamine dihydrochloride (TMPD) were purchased from Sigma-Aldrich.

2.3. Instruments and Apparatus

A vortex mixer (type 37600 mixer, Barnstead/Thermolyne, Dubuque, IA, USA), a centrifuge (Allegra X-22R, Beckman Coulter, Fullerton, CA, USA), a nitrogen evaporator (N-Evap-111, Organomation Associates Inc., Berlin, Germany), and a nitrogen generator (Model 05B, System Instruments Co., Tokyo, Japan) were used to prepare samples. The liquid chromatography–tandem mass spectrometry (LC/MS–MS) apparatus comprised an LC system (Agilent Technologies 1200, Agilent Technologies, Palo Alto, CA, USA) and a mass spectrometer (ABI 4000 QTRAP, Applied Biosystems, Foster City, CA, USA). To determine the levels of residues of chloramphenicol classes, nitrofuran metabolites, and quinolone classes in samples, chromatographic separation was performed in an analytical column (Chromolith Performance RP-18e, 100 mm × 3 mm, Merck, Darmstadt, Germany) and a guard column (Chromolith Guard Column RP-18e, 5 mm × 4.6 mm, Merck). In addition, MG and LMG were separated using a Purospher STAR RP-18 endcapped analytical column (100 mm × 2.1 mm × 2 μm, Merck) and Purospher Star RP-18 endcapped guard column (4 mm × 4 mm × 5 μm, Merck).

2.4. Analysis of LC/MS–MS Conditions

An injection volume of 10 μL was used for determining the levels of chloramphenicol classes, MG, LMG, and quinolone classes and 20 μL for nitrofuran metabolites. Chloramphenicol, thiamphenicol, and florfenicol levels were analyzed through gradient elution by using the A1 eluent (0.1% MeOH) and B1 eluent (100% MeOH). A mobile phase gradient was started at 40% B1 for 1 min at a flow rate of 0.5 mL/min, linearly increased to 90% B1 at 4 min, and subsequently maintained constant until 6 min. Thereafter, it was changed to 40% B1 after 6.1 min and maintained constant until 9 min. Danofloxacin, difloxacin, enrofloxacin, flumequine, and sarafloxacin levels were analyzed through gradient elution by using the A2 eluent (0.1% FA) and B2 eluent (100% MeOH). A mobile phase gradient was started at 10% B1 for 1 min at a flow rate of 0.5 mL/min, linearly increased to 95% B1 at 4 min, and subsequently maintained constant until 8 min. Thereafter, it was changed to 10% B1 after 8.1 min and maintained constant until 9 min. The A3 eluent (0.005 M ammonium acetate in 0.1% FA) and B2 eluent (0.005 M ammonium acetate in MeOH) were used as the mobile phase for nitrofuran metabolite analysis. The mobile phase gradient was started with 30% B2 at a flow rate of 0.3 mL/min, increased linearly to 95% B2 in 4 min, further maintained until 6 min, subsequently changed to 30% B2 after 7 min, and maintained constant until 10 min. The A4 eluent (0.1% FA) and B4 eluent (MeOH) were used as the mobile phase for MG and LMG analyses, respectively. These dyes were separated following the gradient program. The mobile phase gradient was started with 10% B4 for 1 min at a flow rate of 0.5 mL/min, increased from 10% to 95% B4 in 4 min, and maintained constant until 8 min. Finally, B4 was changed to 10% after 8.1 min and maintained constant until 11 min. The MS source conditions in ABI 4000 QTRAP were as follows: ion spray voltage of 4.5–5.5 kV, curtain gas of 15 psi, nebulizer gas of 50 psi, auxiliary gas of 60 psi, and source temperature of 50 °C. MS/MS experiments were conducted in multiple reaction monitoring modes (MRMs) for simultaneous detection of all targets, with two precursor-to-product ion transitions monitored for each analyte. The mass spectrometer was set to detect negative and positive ESI interface modes for chloramphenicol and other veterinary drugs, respectively. Supplementary Table S1 lists the retention times and the precursor and corresponding product ions obtained through MRM detection in LC-amenable veterinary analytes. The dwell time for each MRM transition was set at 5 ms. Analyst software (version 1.4, Applied Biosystems, Foster City, CA, USA) was used for instrument control and data acquisition.

2.5. Preparing the Standard Solutions

In volumetric flasks, stock solutions were prepared that contained pesticide standards or individual veterinary drugs by dissolving each analyte (100 mg) in 100 mL of—depending on the solubility of the analytes—ACN, acetone, or MeOH. All types of stock solution were combined and diluted to 1 mg/L to obtain a working standard mixture. We stored all solutions at −20 °C, and before use, a solution was allowed to adjust to room temperature. With these working standard solutions, serial dilution was performed to prepare a series of calibration standards (dilution range 0.5–500 ng/mL).

2.6. Extraction Procedure and Analysis

To detect residues of chloramphenicol classes, MG, LMG, and quinolone classes, we extracted and cleaned each shrimp sample by using a modification of the veterinary drug residue analysis technique reported by Chang et al. [5] and Smith et al. [8] for aquatic products. Briefly, 2 g of sample was weighed in a propylene centrifuge tube (volume 50 mL) and transferred to a homogenizer containing 100 μL of internal standards (100 ng/mL), 50 μL of TMPD, and 10 mL of ACN. Then, we added 5 mL of n-hexane saturated with CAN to the homogenate, which was shaken in a vortex mixer for 5 min, followed by centrifugation at 4500 rpm for 10 min. We aspirated and subsequently discarded the hexane layer. The ACN extraction layer was collected and dried at 40 °C in a nitrogen evaporator. We re-extracted the remaining tissue pellets using 10 mL of ACN and 5 mL of ACN-saturated n-hexane and then centrifuged them. The first extract was combined with the ACN layer. Subsequently, we evaporated the combined extracts to dryness at 0.5 mL. An additional 0.5 mL of ultrapure water was added, after which the vortex was mixed and then sonicated for 1 min. The reconstituted extracts underwent centrifugation at 4500 rpm for 5 min. Finally, a 0.2 μm polyvinylidene fluoride filter (Whatman, Maidstone, UK) was employed to filter the supernatant layer, and the filtrate was transferred to an autosampler vial prior to being injected into the chromatographic system.

Nitrofuran metabolite extraction from samples was performed through the execution of a TFDA-procedure-based method [9]. Briefly, in a centrifuge tube measuring 50 mL in volume, we fortified 2 g of a sample with 100 μL of internal standards (100 ng/mL), followed by sequentially adding 0.125 M HCl (9 mL) and 50 mM 2-NBA in MeOH (400 μL). Samples were vortex mixed (1 min), followed by overnight incubation (16 h, 37 °C) with gentle shaking in a water bath. In order to neutralize the samples, we added 0.8 M NaOH (1 mL) and 0.1 M dipotassium phosphate buffer (1 mL), and we adjusted the reaction mixture to pH 7.1–7.5. The mixture underwent 1 min of vigorous vortex mixing and was then centrifuged at 3500 rpm for 5 min. After the collection of the supernatant, the remaining tissue pellet was re-extracted using ultrapure water (3 mL), as described earlier in the text, and centrifuged again. The combined extracts were re-extracted using 0.5 g of NaCl and 12 mL of EtOAc with vortex shaking of the samples for 1 min. The reconstituted extracts were again centrifuged for 5 min at 3500 rpm. The solvent was evaporated at 40 °C in a nitrogen evaporator. We reconstituted the resultant dry extract in 1 mL of 50% MeOH, after which it was vortex mixed for 1 min. Subsequently, 1 mL of n-hexane was added to the extracts, which then underwent centrifugation again, as described earlier. The lower layer was collected and filtered (0.2 μm filter membrane). The filtrate was placed in an autosampler vial prior to analysis.

2.7. Assurance and Validation of Quality

We validated the proposed method by calculating the recovery, linearity range, repeatability, and limits of quantification (LOQs) [10,11]. For determining the recovery and repeatability, we spiked blank samples in triplicate by using the following standard mixture of analytes at two concentrations (low and high levels): 1 and 5 ng/g for determination of chloramphenicol classes, dyes, and nitrofuran metabolites; and 5 and 25 ng/g for determination of quinolone classes. Extraction and treatment of the samples followed a previously reported protocol [2,8,9]. The aforementioned recovery validation method was employed to determine the method's repeatability, which was calculated as the percentage of the relative standard deviation (RSD%). The recovery and repeatability (expressed as the percentage of relative standard deviation) of veterinary drugs ranged from 88.67% to 92.35% (repeatability range: 3.79–9.67%) for chloramphenicol classes, 75.21% to 103.31% (repeatability range: 6.72–14.58%) for quinolone classes, 98.81% to 100.31% (repeatability range: 3.58–8.13%) for MG and LMG, and 99.29% to 100.52% (repeatability range: 0.98–5.58%) for nitrofuran metabolites in shrimp samples (Supplementary Table S2). Matrix-matched calibration executed through the use of blank sample extracts and addition of the corresponding amount of working solution (with target compounds at a concentration of 0.5–500 ng/mL) was performed to evaluate the linearity. The calibration curves obtained had high linearity and reproducibility, with the analytical matrix-matched calibration achieving favorable correlation coefficients (R^2 > 0.990). The LOQs were defined as being the concentrations of analyte that yielded peak signals 3× and 10× the intensity of background noise from the chromatogram. The florfenicol, thiamphenicol, chloramphenicol, LMG, MG, and nitrofuran metabolite LOQ was 0.25 ng/mL in shrimp samples. Compared with these chemicals, the LOQ of other veterinary drugs, including danofloxacin, difloxacin, enrofloxacin, flumequine, and sarafloxacin, was 1 ng/g (Supplementary Table S2); concentrations lower than these LOQs indicated that the chemicals and drugs were considered undetectable.

2.8. EDI

To assess the degree to which people are exposed to veterinary drug residues in shrimp, we estimated the EDI from the residual veterinary drug concentrations. The acceptable daily intakes (ADIs) established by the World Health Organization (WHO) and Food and Agriculture Organization of the United Nations (FAO) were employed as points of comparison. The following equation was used to calculate the EDI: EDI (ng/kg/day) = (daily fish consumption [g/day]) × (mean veterinary drug concentration [ng/g])/(human body weight [kg]) [6]. Data regarding Taiwanese citizens' daily seafood consumption (96.9 g for men and 74.2 g for women) were collected from the National Nutrition and Health Survey conducted by the Ministry of Health and Welfare [12]. We considered the mean Taiwanese body weight to be 60 kg [12]. We determined the maximal EDI from the maximum residue concentrations.

3. Results

3.1. Detection Rates and Levels of Veterinary Drugs in Shrimp Samples

In total, 23 whiteleg, 16 grass, and 14 giant river shrimp samples were collected. Chloramphenicol was detected in one grass shrimp, enrofloxacin in one whiteleg shrimp, and flumequine in one whiteleg shrimp and one giant river shrimp (Table 1). These detected veterinary drugs are prohibited by the TFDA for use in shrimp. In all shrimp samples, the predominant residue was flumequine at 3.77% (2/53), followed by chloramphenicol at 1.89% (1/53) and enrofloxacin at 1.89% (1/53). Veterinary drugs were detected in 8.70% (2/23), 6.25% (1/16), and 7.14% (1/14) of the whiteleg, grass, and giant river shrimp samples, respectively. The levels of chloramphenicol and enrofloxacin were 0.29 and 5.62 ng/g in one grass and whiteleg shrimp, respectively. Moreover, flumequine (1.01–1.52 ng/g) was detected in two shrimp samples, namely in one whiteleg and one giant river shrimp. The concentrations of chloramphenicol, enrofloxacin, and flumequine (derived from all samples, including samples with detected and undetected concentrations) were 0.01, 0.11, and 0.05 ng/g, respectively. Overall, 7.55% (4/53) of all shrimp samples contained detectable veterinary drug residues, which indicated the positive ratio of banned residual drugs.

Table 1. Detection levels of banned veterinary drugs in various shrimp samples collected between July 2016 and December 2017.

Shrimp	Surveyed Samples	Violated Targets (No.)	Detected Residues (ng/g)	Average [1] (ng/g, Residues)	Violated Ration [2] (%)
Whiteleg shrimp	23	enrofloxacin (1) flumequine (1)	5.62 1.52	0.24 (enrofloxacin) 0.07 (flumequine)	3.77 (2/53)
Grass shrimp	16	chloramphenicol (1)	0.31	0.02 (chloramphenicol)	1.89 (1/53)
Giant river shrimp	14	flumequine (1)	1.01	0.07 (flumequine)	1.89 (1/53)
Total	53	chloramphenicol (1)	0.31	0.01	
	53	enrofloxacin (1)	5.62	0.11	7.55 (4/53)
	53	Flumequine (2)	1.01–1.52	0.05	

[1] Estimated from all samples, including samples with detected and undetected concentrations. [2] Samples with residual concentrations lower than the LOQ were considered to have undetectable concentrations.

3.2. EDIs of Taiwanese Adults for Veterinary Drug Residues in Shrimp Samples

The Joint FAO/WHO Expert Committee on Food Additives (JECFA) determined the inappropriateness of establishing a chloramphenicol ADI [13]. Therefore, we did not estimate the EDI of chloramphenicol residues in shrimp samples. The EDIs calculated from the average enrofloxacin and flumequine levels were, respectively, 0.14 and 0.06 ng/kg body weight/day for women and 0.18 and 0.08 ng/kg body weight/day for men (Table 2). Regarding the veterinary drug residues in food, the ADIs stipulated by the JECFA's Joint Meeting of the FAO/WHO for enrofloxacin and flumequine are 0.002 and 0.03 mg/kg, respectively [14,15]. As detailed in Table 2, the obtained EDIs were considerably lower than the enrofloxacin and flumequine ADIs recommended by the FAO/WHO. For enrofloxacin and flumequine, the EDIs expressed as a percentage of the ADIs were, respectively, 0.01% and 0.0003% for men and 0.01% and 0.0002% for women. Overall, consumption of shrimp lead to a low risk of dieldrin exposure, with the ADIs lower than 1.0% for men and women.

Table 2. Estimated dietary intake of quinolone residues in Taiwanese adults.

OCPs	EDI (ng/kg Body Weight/Day)		EDI% of ADI		ADI (FAO/WHO) (mg/kg Body Weight/Day)
	Male	Female	Male	Female	
Enrofloxacin	0.18	0.14	0.01	0.01	0.002
Flumequine	0.08	0.06	0.0003	0.002	0.03

4. Discussion

In the present study, 14 residual veterinary drugs, namely three chloramphenicol classes (chloramphenicol, florfenicol, and thiamphenicol), five quinolone classes (danofloxacin, difloxacin, enrofloxacin, flumequine, and sarafloxacin), MG, LMG, and four nitrofuran metabolites (AMOZ, AOZ, AH, and SC), were analyzed in 52 shrimp samples collected from aquaculture areas in Taiwan. To validate the presence of these compounds in samples, we evaluated the mean recovery (as a measure of trueness), linearity, sensitivity, and repeatability (as a measure of precision) according to EU guidelines (SANCO/12495/2011) [10]. Because chloramphenicol, nitrofuran metabolites, MG, and LMG are banned from use in edible animals and danofloxacin, difloxacin, enrofloxacin, flumequine, and sarafloxacin are banned from use in decapods, the TFDA does not recommend MRLs in shrimp.

Therefore, the residues of these banned compounds in shrimp were detected and sufficiently indicated the degree of legal compliance regarding the use of these products.

The Commission Decision 2002/657/EC criteria for evaluation of veterinary drug residues in animals and animal products are stipulated on the basis of mass spectrometry at numerous identification points (IPs) [16]. Source conditions were optimized to obtain 1.5 IP from product ions and one IP from precursor ions for each compound. In general, obtaining four IPs at the lowest level is required for analyzing banned compounds. In the present study, veterinary drugs were analyzed in the MRM mode by monitoring three different ions (one precursor and two fragment ions). Using this approach, we achieved four IPs (one IP from a single precursor ion and three IPs from two fragment ions), as mandated by the aforementioned guidelines. Our analysis method successfully identified the residues of MG, LMG, chloramphenicol classes, quinolone classes, and nitrofuran metabolites.

The analytical extraction method for aquatic samples was designed by Smith et al. [8]. In this method, ACN and hexane are used to extract samples for simultaneously screening multiple classes of drug residues, including macrolides, β-lactam antibiotics, dyes, quinolones, tetracyclines, and antimycotic imidazoles. Moreover, other extraction methods have been reported for analyzing chloramphenicol [17], MG, and LMG [18]. In addition, we applied this method for the extraction of chloramphenicol, MG, and LMG residues in bivalve samples [5]. The method used herein was developed for the simultaneous detection of chloramphenicol classes, quinone classes, MG, and LMG in shrimp samples. This is the most efficient and energy-conservative method for veterinary drug extraction. However, the same method could not be employed to analyze the nitrofuran metabolite residues in aquatic samples; because of their chemical structural characteristics, nitrofuran metabolites in food samples were extracted using 2-NBA for derivatization [19].

To validate the shrimp sample analysis method, as recommended by the TFDA [11], the acceptable recovery rate had to be 70–120% with RSD < 15% for chemical residues in food matrixes detected in the 0.1–1 mg/kg range; 70–120% with RSD < 20% for those detected in the 0.01–0.1 mg/kg range; 60%–125% with RSD < 30% for those detected in the 0.001–0.01 mg/kg range; and 50–125% with RSD < 35% for those detected within ≤0.001 mg/kg. According to our results, veterinary drug residues detected within ≤0.001 mg/kg and 0.001–0.01 mg/kg ranges demonstrated a recovery rate of 80–120% with an RSD of <10% and a recovery rate of 70–120% with an RSD of <15%, respectively. The TFDA also recommends various LOQs, including 0.3 ng/g in chloramphenicol; 5 ng/g in florfenicol and thiamphenicol; 10 ng/g in quinolone classes; 0.5 ng/g in MG and LMG; and 1 ng/g in nitrofuran metabolites, for aquatic food for the assessment of veterinary drug residues in seafoods [20]. Compared with the LOQs recommended by the TFDA, the LOQs obtained using our analytical method were lower and can be employed to detect trace veterinary drug residues. Therefore, the analytical methods employed herein conform to the recommendations of the TFDA.

The regulations entitled Tolerances for Residues of Veterinary Drugs in Food, established by the Ministry of Health and Welfare of Taiwan, state that nitrofuran metabolites, chloramphenicol, MG, and LMG are banned from use in shrimp culturing because of concerns that pertain to mutagenicity and carcinogenicity [5]. In addition, food-producing animals and products containing these chemicals exported by third-world countries are prohibited in Japan and the European Union, the major importers of Taiwanese marine products. Based on methodologies available for detecting banned compounds in edible products, the Department of Health of Taiwan [21] and EU Commission [22] have both established a maximum residual permissible limit (MRPL) of 1 ng/g for each nitrofuran metabolite in aquaculture, marine, and poultry meat products. Furthermore, the EU Commission stipulates an MRPL of 0.3 ng/g for chloramphenicol and 2 ng/g for MG plus LMG [16] in all food products of animal origin to ensure that customers worldwide are given the same level of protection. According to the aforementioned guidelines, the LOQs of our methods executed for identifying the levels of chloramphenicol, dye, and nitrofuran metabolite residues in shrimp meet the MRPL.

The chromatography–mass spectrometry screening of carcinogenic antimicrobials—such as nitrofuran metabolites, chloramphenicol, MG, and LMG—in 53 shrimp samples demonstrated a positive result, with the chloramphenicol concentration being greater than the MRPL of chloramphenicol set by the EU Commission (0.31 ng/g in a grass shrimp sample). Administering chloramphenicol to food-producing animals is banned in Taiwan. Some aquaculture farmers use chloramphenicol regardless, however, because it is a broad-spectrum, inexpensive, and readily available antibiotic [23]. Mixed breeding has caused chloramphenicol to be employed for the prevention as well as treatment of infectious diseases in shrimp. In the present study, the proportion of positive identification of banned veterinary drugs was 1.89% (1/53). In the analysis of cultured shrimp in Bangladesh, the detection of chloramphenicol and nitrofuran metabolite residues revealed a violation ratio of 8.37% (118/1409) in 2008, 8.16% (182/2230) in 2009, and 5.81% (122/2098) in 2009 [24]. In addition, in the Canadian Total Diet Study from 1994 to 2004, the detection of MG, LMG, and nitrofuran metabolite residues revealed a violation ratio of 20.0% (6/30) [25]. In Ireland, exposure to nitrofuran residues was assessed from 2009 to 2010, which revealed a violation ratio of 5.68% (5/88) in the detection of SEM residues [26]. However, in the aforementioned reports, only two classes of nitrofuran metabolites and chloramphenicol or three classes of MG, LMG, and nitrofuran metabolites were detected. Our present findings differ from those of TFDA surveys. In reports in recent years, the violation ratio of banned veterinary drug residues in shrimp samples was 0% in 2013 (0/20) [27] and 0% in 2014 (0/20) [28]. These differences are partially accounted for by varying sample sizes. In addition, the samples collected in this study were obtained from shrimp production areas in Taiwan, whereas the samples collected by the TFDA may have been imported shrimp. Therefore, several categories of banned veterinary drugs in Taiwanese shrimp were appropriately detected in the present study.

In the present investigation, quinolone residues (3/53) were detected with a higher violation ratio than chloramphenicol (1/53) in aquaculture shrimp. Our study revealed that quinolones were the predominant compounds in the aquacultured shrimp samples in Taiwan, which was similar to the results of a survey conducted in China [29], Vietnam [30,31], and Thailand [31] following intensive use in aquaculture to treat bacterial infections, which polluted aquatic habitats and had adverse effects on the health of freshwater organisms. Quinolones were detected in 8.70% (2/23) and 7.14% (1/14) of whiteleg and giant river shrimp samples, respectively. In all shrimp samples, the predominant residue was flumequine at 3.77% (2/53), followed by enrofloxacin at 1.89% (1/53). The results of the present study are similar to those of the survey conducted by the TFDA. Compared with a report of the TFDA in 2012, the violation ratio of quinolone residues in shrimp samples was 4.0% in 25 samples, which was positive with flumequine at 21.0 ng/g in one shrimp sample [32]. From the data available, we concluded that flumequine continues to be used as a growth promoter and prophylactic agent in aquatic products because of its affordability and effectiveness. Other surveys in Asia [31,33] have reported that flumequine has been the most widely used synthetic antibiotic in aquaculture, especially because of its relative stability to resist bacterial degradation in water and sediments. In addition, flumequine residues were detected in trace amounts; only a concentration of 1.01–1.52 ng/g or higher triggers action by the TFDA (withdrawal of the product and issuance of an alert). The results of the surveys reviewed herein indicate that the Taiwanese population is exposed to trace amounts of flumequine that do not pose an immediate risk to health through the consumption of shrimp. Therefore, Taiwan's regulatory authorities and producers should continually monitor aquatic products and prevent sources of contamination, ensuring the chemical safety of commercially available foods.

Parameter guidelines indicate how the risk to organisms such as humans can be assessed by stipulating criteria related to the ADI, hazard quotients, provisional tolerable weekly intake, and excess cancer risk [6,34,35]. Guidelines for the ADI, such as those formulated by the FAO and WHO, facilitate the assessment of risks to organisms, including humans [6]. The ADI is a single value, however, and eating habit and consumption rate differences are not considered in its calculation [36,37]. The JECFA [38] and US EPA [39] have proposed a new and highly accurate measure for the estimation of chronic dietary intake: the EDI. In this study, we concluded that the ADI was not exceeded by the corresponding daily exposure. Because few residual quinolones were discovered, the estimated EDI revealed that consumption of the investigated shrimp would result in considerably less dietary intake of enrofloxacin and flumequine in the Taiwanese population than that stipulated by the ADI. Furthermore, when assessed against the ADIs, the EDIs calculated in this study indicated no risk to health due to shrimp consumption. The EDIs were lower than 1% of the ADIs in this study, indicating negligible risk [6,38]. Thus, the levels of quinolone in Taiwanese food products can be concluded to not negatively affect health. Because of the potentially adverse effects of antibiotics on health and aquatic environments, the impact of these pollutants must be urgently evaluated further.

5. Conclusions

In the present study, we analyzed the residues of chloramphenicol, florfenicol, thiamphenicol, MG, LMG, nitrofuran metabolites, danofloxacin, difloxacin, enrofloxacin, flumequine, and sarafloxacinthe in shrimp samples; methods used were validated according to the EU criteria and complied with the MRPLs established by the EU and TFDA. The residues of banned veterinary drugs chloramphenicol and quinolone, with no MRL recommended, were detected in 53 shrimp samples. We observed that one shrimp sample contained chloramphenicol, one shrimp sample contained enrofloxacin, and two shrimp samples contained flumequine. Notably, only trace amounts of all residues were discovered, indicating no immediate risk to health because the EDIs were considerably lower than the FAO/WHO-defined ADIs. Enrofloxacin and flumequine contamination following shrimp consumption in Taiwan appears to present a negligible threat to human health. However, the concern regarding pharmaceuticals and their adverse effects on the environment and human health is increasing, and a background information system on the consumption of veterinary antibiotics through shrimp must be established and improved, thus providing a monitoring and management framework. The health and agricultural authorities can use the present study findings as a valuable reference when improving contaminant regulation in aquaculture.

Supplementary Materials: The following are available online at http://www.mdpi.com/2076-3417/9/12/2463/s1, Supplementary Table S1: Retention time and MS/MS fragmentation conditions for veterinary drugs and their corresponding internal standards, Supplementary Table S2: Recovery, repeatability, and LOQ of veterinary drugs spiked into whiteleg shrimp.

Author Contributions: M.-Y.T. and C.-F.L. conceived the idea and performed experiments. W.-C.Y., C.-T.L., and K.-H.H. assisted in recombinant construction. G.-R.C. wrote, reviewed, and edited the manuscript. M.-Y.T. and C.-F.L. contributed equally to this work.

Funding: This research received no external funding.

Acknowledgments: This study was supported by the Ministry of Science and Technology (Taiwan) (MOST 107-2313-B-415-012) and, in part, by the Taichung Veterans General Hospital (Taiwan) and National Chung-Hsing University (Taiwan) (TCVGH-NCHU-10776013). This manuscript was edited by Wallace Academic Editing.

Conflicts of Interest: The authors declare no conflict of interest.

References

1. Chang, G.R. Surveys on banned veterinary drugs residues in marine bivalves and gastropods in Taiwan between 2010 and 2015: A mini review. *J. Aquat. Pollut. Toxicol.* **2017**, *1*, 1–5.
2. Liao, I.C. Aquaculture practices in Taiwan and its visions. *J. Fish. Soc. Taiwan* **2005**, *32*, 193–206.
3. Fisheries Agency, Council of Agriculture, Taiwan. 2017 Fisheries Statistical Yearbook. 2018. Available online: https://www.fa.gov.tw/cht/PublicationsFishYear/ (accessed on 5 February 2019).
4. Chen, C.L.; Qiu, G.H. The long and bumpy journey Taiwan's aquaculture development and management. *Mar. Policy* **2014**, *48*, 152–161. [CrossRef]
5. Chang, G.R.; Chen, H.S.; Lin, F.Y. Analysis of banned veterinary drugs and herbicide residues in shellfish by liquid chromatography-tandem mass spectrometry (LC/MS/MS) and gas chromatography-tandem mass spectrometry (GC/MS/MS). *Mar. Pollut. Bull.* **2016**, *113*, 579–584. [CrossRef] [PubMed]
6. Chang, G.R. Persistent organochlorine pesticides in aquatic environments and fishes in Taiwan and their risk assessment. *Environ. Sci. Pollut. Res. Int.* **2018**, *25*, 7699–7708. [CrossRef] [PubMed]

7. Song, K.H.; Breslin, V.T. Accumulation and transport of sediment metals by the vertically migrating opossum shrimp, Mysis relicta. *J. Gt. Lakes Res.* **1999**, *25*, 429–442. [CrossRef]

8. Smith, S.; Gieseker, C.; Reimschuessel, R.; Decker, C.S.; Carson, M.C. Simultaneous screening and confirmation of multiple classes of drug residues in fish by liquid chromatography-ion trap mass spectrometry. *J. Chromatogr. A* **2009**, *1216*, 8224–8232. [CrossRef] [PubMed]

9. Department of Health, Executive Yuan, Taiwan. *Method of Test for Veterinary Drug Residues in Foods-Test of Nitrofuran Metabolites*; Announcement No. 1021950758; Department of Health, Executive Yuan, Taiwan: Taipei, Taiwan, 2013.

10. European Union. *Method Validation and Quality Control Procedures for Pesticides Residues Analysis in Food and Feed*; Document No SANCO/12495/2011; European Union: Brussels, Belgium, 2011.

11. Shen, Y.R.; Cheng, M.W.; Wu, B.S.; Yang, K.C.; Chang, Y.H.; Tseng, S.H.; Kao, Y.M.; Chiueh, L.C.; Shih, D.Y.C. Developement of the method for analysis of multiple pesticide residues in animal matrices by QuEChERS method. Taiwanese. *J. Agric. Chem. Food Sci.* **2013**, *51*, 148–160.

12. Wu, S.J.; Chang, Y.H.; Fang, C.W.; Pan, W.H. Food sources of weight, calories, and three macro-nutrients-NAHSIT 1993–1996. *Nutr. Sci. J.* **1999**, *24*, 41–58.

13. Nicolich, R.S.; Werneck-Barroso, E.; Marques, M.A.S. Food safety evaluation: Detection and confirmation of chloramphenicol in milk by high performance liquid chromatography-tandem mass spectrometry. *Anal. Chim. Acta* **2006**, *565*, 97–102. [CrossRef]

14. He, X.T.; Deng, M.C.; Wang, Q.; Yang, Y.T.; Yang, Y.F.; Nie, X.P. Residues and health risk assessment of quinolones and sulfonamides in cultured fish from Pearl River Delta, China. *Aquaculture* **2016**, *458*, 38–46. [CrossRef]

15. Itoh, T.; Mitsumori, K.; Kawaguchi, S.; Sasaki, Y.F. Genotoxic potential of quinolone antimicrobials in the in vitro comet assay and micronucleus test. *Mutat. Res.* **2006**, *603*, 135–144. [CrossRef] [PubMed]

16. European Commission. Directive 2002/657/EC concerning the performance of analytical methods and the interpretation of results. *Off. J. Eur. Commun.* **2002**, *L221*, 8–36.

17. Takinoa, M.; Daishimab, S.; Nakaharac, T. Determination of chloramphenicol residues in fish meats by liquid chromatography–atmospheric pressure photoionization mass spectrometry. *J. Chromatogr. A* **2003**, *1011*, 67–75. [CrossRef]

18. Lee, K.C.; Wu, J.L.; Cai, Z. Determination of malachite green and leucomalachite green in edible goldfish muscle by liquid chromatography-ion trap mass spectrometry. *J. Chromatogr. B Anal. Technol. Biomed. Life Sci.* **2006**, *843*, 247–251. [CrossRef] [PubMed]

19. Radovnikovic, A.; Moloney, M.; Byrne, P.; Danaher, M. Detection of banned nitrofuran metabolites in animal plasma samples using UHPLC-MS/MS. *J. Chromatogr. B Anal. Technol. Biomed. Life Sci.* **2011**, *879*, 159–166. [CrossRef] [PubMed]

20. Fu, H.P.; Chen, C.M.; Hang, S.F.; Lin, Y.Y.; Lin, Y.R.; Chen, T.L. Post-market surveillance study on veterinary drug residues in poultry, livestock and aquatic products in 2017. *Ann. Rep. Food Drug Res.* **2018**, *9*, 115–124.

21. Ministry of Health and Welfare, Executive Yuan, Taiwan. *Tolerances for Residues of Veterinary Drugs*; MOHW Food No. 1041303515; Ministry of Health and Welfare, Executive Yuan, Taiwan: Taipei, Taiwan, 2015.

22. European Commission. Decision 2003/181/EC amending decision 2002/657/EC as regards the setting of minimum performance limits (MRPLs) for certain residues in food animal origin. *Off. J. Eur. Commun.* **2013**, *L71*, 17.

23. Defoirdt, T.; Sorgeloos, P.; Bossier, P. Alternatives to antibiotics for the control of bacterial disease in aquaculture. *Curr. Opin. Microbiol.* **2011**, *14*, 251–258. [CrossRef] [PubMed]

24. Hassan, M.N.; Rahman, M.; Hossain, M.B.; Hossain, M.M.; Mendes, R.; Nowsad, A.A.K.M. Monitoring the presence of chloramphenicol and nitrofuran metabolites in cultured prawn, shrimp and feed in the Southwest coastal region of Bangladesh. *Egypt. J. Aquat. Res.* **2013**, *39*, 51–58. [CrossRef]

25. Tittlemier, S.A.; Riet, J.; Burns, G.; Potter, R.; Murphy, C.; Rourke, W.; Pearce, H.; Dufresne, G. Analysis of veterinary drug residues in fish and shrimp composites collected during the Canadian Total Diet Study, 1993–2004. *Food Addit. Contam.* **2007**, *24*, 14–20. [CrossRef]

26. Radovnikovic, A.; Conroy, E.R.; Gibney, M.; O'Mahony, J.; Danaher, M. Residue analyses and exposure assessment of the Irish population to nitrofuran metabolites from different food commodities in 2009–2010. *Food Addit. Contam. A* **2013**, *30*, 1858–1869. [CrossRef] [PubMed]

27. Fu, H.P.; Kuo, H.W.; Shih, C.C.; Lin, H.C.; Lin, K.H.; Lin, Y.R.; Chou, H.K.; Shyu, J.F.; Pan, J.Q.; Hsu, C.K.; et al. 2013 Post-market survey on veterinary drug residues in livestock and aquatic products. *Ann. Rep. Food Drug Res.* **2014**, *5*, 81–91.

28. Fu, H.P.; Lin, Y.P.; Su, H.C.; Wang, T.S.; Hsu, C.H.; Liu, F.M.; Feng, R.L.; Hsu, C.K.; Liu, L.W.; Tzeng, G.S.; et al. 2014 Post-market survey on veterinary drug residues in livestock and aquatic products. *Ann. Rep. Food Drug Res.* **2015**, *6*, 67–75.

29. Zhao, S.; Jiang, H.; Li, X.; Mi, T.; Li, C.; Shen, J. Simultaneous determination of trace levels of 10 quinolones in swine, chicken, and shrimp muscle tissues using HPLC with programmable fluorescence detection. *J. Agric. Food Chem.* **2007**, *55*, 3829–3834. [CrossRef] [PubMed]

30. Pham, D.K.; Chu, J.; Do, N.T.; Brose, F.; Degand, G.; Delahaut, P.; De Pauw, E.; Douny, C.; Nguyen, K.V.; Vu, T.D.; et al. Monitoring antibiotic use and residue in freshwater aquaculture for domestic use in vietnam. *Ecohealth* **2015**, *12*, 480–489. [CrossRef] [PubMed]

31. Takasu, H.; Suzuki, S.; Reungsang, A.; Pham, H.V. Fluoroquinolone (FQ) contamination does not correlate with occurrence of FQ-resistant bacteria in aquatic environments of Vietnam and Thailand. *Microbes Environ.* **2011**, *26*, 135–143. [CrossRef]

32. Fu, H.P.; Kuo, H.W.; Shih, C.C.; Lin, H.C.; Lin, Y.R.; Chou, H.K.; Shyu, J.F.; Pan, J.Q.; Hsu, C.K.; Liu, L.W.; et al. 2012 Post-market survey on veterinary drug residues in livestock and aquatic products. *Ann. Rep. Food Drug Res.* **2013**, *4*, 38–46.

33. Thuy, H.T.; Nga le, P.; Loan, T.T. Antibiotic contaminants in coastal wetlands from Vietnamese shrimp farming. *Environ. Sci. Pollut. Res. Int.* **2011**, *18*, 835–841. [CrossRef]

34. Gu, Y.G.; Lin, Q.; Huang, H.H.; Wang, L.G.; Ning, J.J.; Du, F.Y. Heavy metals in fish tissues/stomach contents in four marine wild commercially valuable fish species from the western continental shelf of South China Sea. *Mar. Pollut. Bull.* **2017**, *114*, 1125–1129. [CrossRef]

35. Ke, C.L.; Gu, Y.G.; Liu, Q.; Li, L.D.; Huang, H.H.; Cai, N.; Sun, Z.W. Polycyclic aromatic hydrocarbons (PAHs) in wild marine organisms from South China Sea: Occurrence, sources, and human health implications. *Mar. Pollut. Bull.* **2017**, *117*, 507–511. [CrossRef] [PubMed]

36. Yang, N.; Matsuda, M.; Kawano, M.; Wakimoto, T. PCBs and organochlorine pesticides (OCPs) in edible fish and shellfish from China. *Chemosphere* **2006**, *63*, 1342–1352. [CrossRef] [PubMed]

37. Li, X.; Nie, X.P.; Pan, D.B.; Li, G.Y. Analysis of PAEs in muscle tissue of freshwater fish from fishponds in Pearl River Delta. *J. Environ. Health* **2008**, *25*, 202–204.

38. Vragović, N.; Bazulić, D.; Njari, B. Risk assessment of streptomycin and tetracycline residues in meat and milk on Croatian market. *Food Chem. Toxicol.* **2011**, *49*, 352–355. [CrossRef] [PubMed]

39. Zhang, G.; Pan, Z.; Bai, A.; Li, J.; Li, X. Distribution and bioaccumulation of organochlorine pesticides (OCPs) in food web of Nansi Lake, China. *Environ. Monit. Assess.* **2014**, *186*, 2039–2051. [CrossRef] [PubMed]

Article

Detection and Treatment Methods for Perfluorinated Compounds in Wastewater Treatment Plants

Shun-hwa Lee [1], Yeon-jung Cho [1], Miran Lee [2] and Byung-Dae Lee [3,*]

[1] Department of Environmental Engineering, Yeungnam University, Gyeongsan 38541, Korea; leesh@yu.ac.kr (S.-h.L.); dkssud4878@naver.com (Y.-j.C.)
[2] Daisung Green Tech, Seongnam, Gyenggi 13216, Korea; dr88@chol.com
[3] Department of Health Management, Uiduk University, Gyeongju 38004, Korea
* Correspondence: bdlee@uu.ac.kr; Tel.: +82-54-760-1702; Fax: +82-54-760-1179

Received: 8 May 2019; Accepted: 13 June 2019; Published: 19 June 2019

Abstract: We surveyed the variation in perfluorinated compound (PFC) concentrations entering urban wastewater treatment plants and then designed an optimal PFCs treatment method based on a pilot test. The PFCs influent concentration was found to be affected by the types of industries and operating rate. The concentration of PFCs in the wastewater treatment effluent was slightly lower than that of the influent. Thus, PFCs had not been adequately removed by the existing biological treatments. The pilot test results showed that about 10% of PFCs was removed by coagulation and precipitation, and the ozone and chlorine test showed that few, if any, PFCs were removed regardless of the oxidant dose. The activated carbon adsorption test showed that the removal significantly increased with empty bed contact time, with about a 60% removal in five minutes and over 90% removal in over 15 minutes. Therefore, a more stable and higher PFCs removal would result from continuous oxidation processes, such as ozone and adsorption processes involving activated carbon, rather than a single biological treatment.

Keywords: perfluorinated compounds; coagulation; ozone; chlorination; activated carbon

---

1. Introduction

Perfluorinated compounds (PFCs) do not easily degrade biologically in the natural environment due to their extremely stable covalent bonds. Artificially synthesized and produced, PFCs are used in manufacturing various household goods and are detected consistently in water [1]. There has been much research interest in PFCs as they have been reported to affect the natural water system, causing disturbances to the ecosystem [1,2]. Since the mid-2000s, the European Union (EU), Canada, and the USA, among others, have started regulating PFCs. At the fourth Stockholm Treaty Conference of the Parties in May 2009, it was agreed that some of these PFCs would be included in the target compounds of the treaty [3–5]. The most well-known PFCs are used in industry and broader daily life for waterproof materials, lubricants, paint, ink, paper, fiber, carpet, ovens, cooking equipment, electronic products, packaging materials, metal coating, cleaning products, semi-conductors, and firefighting products [2]. However, since many household products use PFCs as additives, it is difficult to calculate their amount of usage and the amounts released into the environment. Whereas the residual concentration of PFCs in the environment is small, at the ng/L (ppt) level or below, the effects on the ecosystem cannot be disregarded if the ecosystem then is exposed long-term to PFCs [6–8].

The pathways of PFCs to the environment are numerous; most of them enter via wastewater treatment plants. However, little has been known about the mechanisms through which these persistent and toxic PFCs are removed. Therefore, it is difficult to improve the removal of the small amount of PFCs with existing wastewater treatment methods [9,10].

In this study, we aimed to investigate PFCs concentration in the influent and effluent of an actual wastewater treatment plant, which is the source of perfluorocarbons in large industrial complexes and highly populated areas. The distribution properties of PFCs were investigated from the influent wastewater treatment plant and an optimal treatment plan was designed through a pilot test, which may eventually serve as data for future surface water management plans.

2. Research Content and Method

2.1. Current Status of Wastewater Treatment Plants

2.1.1. Survey Area

The major business types in the survey area are fabric manufacturing, metal, coating, and rubber, and the fabric manufacturers produce the largest amounts of PFCs emissions [9]. The public wastewater treatment plant implements the anaerobic-anoxic-oxic (A2/O) method, and the reaction tank consists of an anaerobic tank, an anoxic tank, and an aerobic tank, as well as internal and external returners. The treatment efficiency for organic compounds like BOD and SS is over 90%, that for TN is between 40% and 70%, and that for TP is 60%. Figure 1 shows the process flowchart of the wastewater treatment plants in city K.

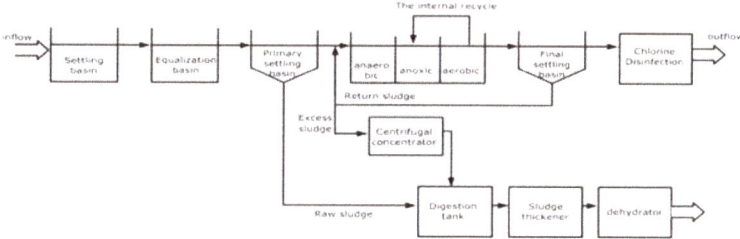

Figure 1. The process flowchart of the wastewater treatment plants in city K.

2.1.2. Analysis of PFCs

The PFCs analysis was conducted on five compounds: perfluorooctanoic acid (PFOA), perfluoro-n-pentanoic acid (PFPeA), perfluorohexane sulfonate (PFHxS), perfluorohexanoic acid (PFHxA), and perfluorononanoic acid (PFNA). The pre-treatment and analysis conditions are provided in Table 1. The pre-treatment for the analysis of PFCs involves the use of solid-phase extraction (SPE) to complete a four-stage solid phase extraction, and the conditioning stage uses methanol as the solvent. The loading and washing stages used 10.0 mL of de-ionized water, and the elution stage used 2.0 mL of methanol. After the solid phase extraction stage, the concentration stage used 40 °C nitrogen gas to concentrate the PFCs so that their final volume was reduced to 500 μL. After the concentration stage, an analysis was conducted with LC-MS [11].

Table 1. Analytical condition of perfluorinated compounds (LC: Liquid chromatography; MS: mass spectrometer).

Parameter	Conditions
LC	Waters, ACQUITY, HPLC
MS	SCIEX, API-4000, Q-Trap
Column	Waters, BEH C18, 2.1 × 50 mm, 1.7 µm
Mobile phase	A: 10 mM ammonium acetate in water
	B: Acetonitrile
Gradient	Time (min)
	Solvent B (%)
Flow rate	270 µL/min
Injection volume	5 µL
Column temp.	40 °C
Ionization mode	Negative
Curtain Gas	40
Collision Gas	8
Ion Spray Voltage	−4500 V

2.1.3. Pilot Test

To determine the concentration change by coagulation-sedimentation, we injected the coagulator using Alum at regular intervals of 10, 20, 30, 40, and 50 mg/L. After fixing the pH to 7, we conducted rapid spinning (67 rpm) for one minute, slow spinning (31 rpm) for 10 minutes, and then sedimentation for 30 minutes. The supernatant was then sampled and analyzed. The PFCs removal efficiency by ozone oxidation was determined during an ozone contact test. The ozone contact test was performed by determining the ozone dose by potassium iodide titration and injecting the ozone to the water in the Ozone Demand Flask in rising concentrations of 5, 10, 15, 20, 25, and 30 mg O_3/L. The volume of the Demand Flask was 1.5 L, and after injecting ozone, the Demand Flask was sufficiently shaken for 20 minutes to maximize the ozone contract efficiency before carrying out the analysis. By measuring the effective chlorine of the hypochlorous acid sodium solution in the inflow water to the treatment plant, the corresponding chlorine dose was calculated and injected. The injection doses of chlorine were 5, 10, 15, and 20 mg/L at regular intervals, based on the guidelines for the sewage treatment facilities. For the activated carbon adsorption of PFCs, we placed clearly washed and dried granular activated carbon into a column with an internal diameter of 20 mm and length of 350 mm. The influent wastewater was injected into this column using a controlled volume pump, and the treated sample was then analyzed. For the test conditions, the empty bed contact time (EBCT) was varied to 5, 10, and 15 minutes.

3. Results and Discussion

3.1. Variation in Amount of Inflow Water

Figure 2 sows the amount of influent flow to the treatment plant. The daily average influent flow was 27,335 m³/day, 40,948 m³/day at the highest level, and 19,506 m³/day at the lowest level. Due to the precipitation in August 2017, the inflow water between the August 13 and 17 was the highest. The amount of influent flow partially changes due to business operational rates.

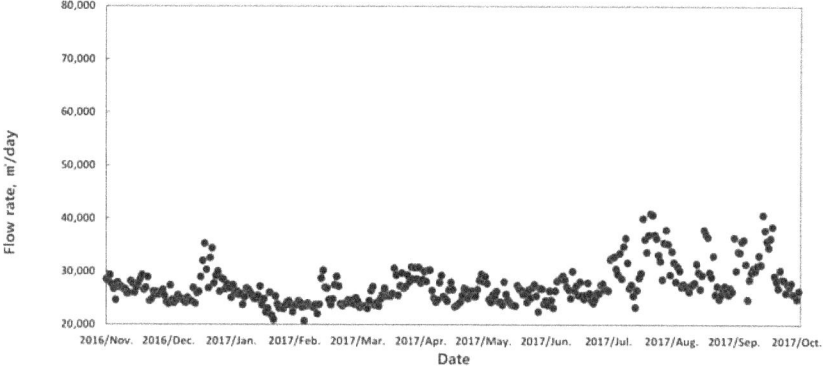

Figure 2. Flow variations of a wastewater treatment plant in city K.

3.2. Monthly PFCs Concentration Change

The average PFOA in the influent was 11.7 ng/L, and that in the effluent was 11.0 ng/L. The PFOA removal was very low. The average PFHxA in the influent was 11.9 ng/L and that in the effluent was 11.1 ng/L, showing the minimal removal. The average PFNA in the influent was 4.1 ng/L, but that in the effluent was 2.9 ng/L, showing an average PFNA removal of 27.2%. The average PFPeA in the influent was 3.2 ng/L and the effluent was 2.7 ng/L, which is 7.8% removal. The average PFHxS in the influent was 1.0 ng/L, and that in the effluent was 0.8 ng/L.

The concentration distribution of these substances tended to be somewhat higher in the dry winter season. PFHxS is frequently used as a fire extinguisher material and so its concentration tended to be higher in the fall and winter when more fires occur. In addition, PFCs are shown to be largely affected by business operational ratios, and thus, for efficient management, a detailed and continuous emissions survey would be required in industry-concentrated regions. Finally, whereas some compounds were removed by the existing treatments, most of the compounds were shown to be poorly treated.

3.3. Correlation of PFCs between Cities

The correlation of PFCs with other cities with similar wastewater characteristics was compared. As shown in Table 2, correlational coefficients between 0.000 and 0.200 were considered to have no relation and are marked as xx, whereas those between 0.210 and 0.400 were considered to be low and are marked as x. A correlational coefficient between 0.410 and 0.600 was considered to have some relation and is marked as △.

Table 2. Comparison of correlation coefficients of PFCs with other cities with similar populations and industries (xx: 0-0.2; x: 0.21-0.4; △: 0.41-0.6; PFOA: perfluorooctanoic acid; PFPeA: perfluoro-n-pentanoic acid; PFHxS: perfluorohexane sulfonate; PFHxA: perfluorohexanoic acid; PFNA: perfluorononanoic acid).

R^2			Other Cities							
Compounds	PFOA		PFHxA		PFNA		PFPeA		PFHxS	
PFOA	1	-	0.050	xx	0.013	xx	0.141	xx	0.486	△
PFHxA	-	-	1	-	0.516	△	0.583	△	0.090	xx
PFNA	-	-	-	-	1	-	0.160	xx	0.019	xx
PFPeA	-	-	-	-	-	-	1	-	0.098	xx
PFHxS	-	-	-	-	-	-	-	-	1	-

The correlation coefficient between PFOA and PFHxS was 0.486, showing some correlation. That between PFHxA and PFNA or PFPeA was 0.516 or 0.583, respectively, showing some correlation. All the other compounds had low or little correlation. The correlation among PFCs showed that PFOA, PFHxA, and PFNA have the –COOH group, so that were highly correlated with PFCs with the same reaction group [3]. Ahrens et al. also showed that the correlation coefficient between PFO and PFNA was highest at 0.752 [12]. However, So et al. reported that whereas PFCs in the –COOH group had a relatively higher correlation, the correlation coefficient between PFOA and PFNA was low at 0.375 [13]. This occurred because there are various emission sources for PFCs, and the outflow water from the wastewater treatment plants are known to be key point pollution sources [3]. Therefore, effluent flow to the water system reflects regional characteristics and determines the correlation of the substances in actual streams, producing these regional differences [3].

The correlational study among PFCs was significant in determining the concentration level of PFCs through the concentration of some PFCs among those with high correlation. However, those data are still lacking in representability; therefore, further research is necessary.

3.4. PFCs Removal Efficiency Through Pilot Test

We conducted a pilot test on the influent to the treatment plant to determine the removal efficiency. In particular, we examined the changes in the removal by coagulation-sedimentation, ozone oxidization, chlorine injection, and activated carbon adsorption.

3.4.1. Removal Variation by Coagulation-Sedimentation

As shown in Figure 3, the PFCs in the coagulation-sedimentation test showed a phased concentration decrease. However, overall, the removal did not increase despite the additional injection of the coagulator of more than 40–50 mg/L. When 40–50 mg/L of coagulator was injected, the removal of PFPeA was highest at 12.3%, followed by PFHxS at 11.5%, PFNA at 10.9%, PFHxA at 8.9%, and PFOA at 6.4%.

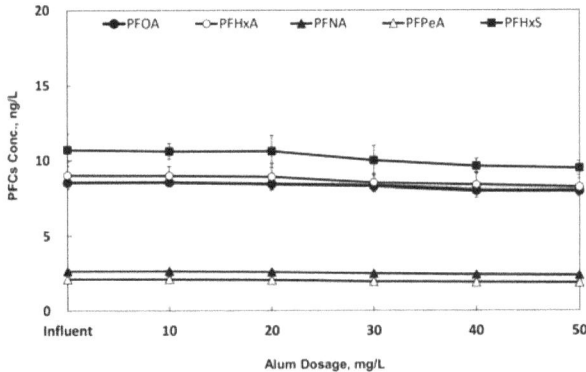

Figure 3. Variations in PFCs concentrations with Alum dosages.

The overall removal of PFCs was about 10%, demonstrating that they were somewhat removed by coagulation-sedimentation treatment. However, the effect was shown to be negligible and that it would be limited in removing non-degradable PFCs.

3.4.2. Removal Variation by Ozone Oxidation

The ozone contact test results in Figure 4 show that, in all compounds from PFOA to PFHxA, PFNA, PFPeA, and PFHxS, the removal was 5% or lower. Thus, regardless of the change in the ozone dose, the removal was negligible. Previous research showed that the PFOA removal was between 10% and 20%, but the removal by ozone oxidization was difficult [9].

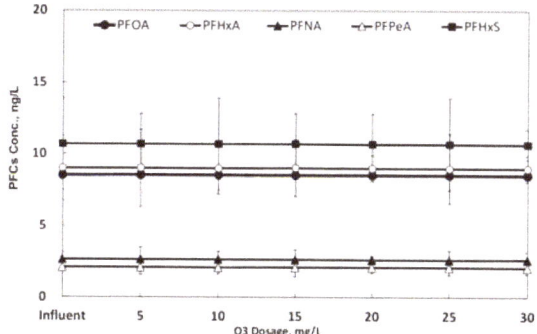

Figure 4. Variations of PFCs concentration by O_3 dosages.

Due to the highly stable structure, PFCs are difficult to remove through existing biochemical or oxidization processes. We also determined that PFCs removal by ozone oxidization would be difficult. Therefore, to remove them effectively, other additional treatment methods are required.

3.4.3. Removal Change by Chlorine Injection

As shown in Figure 5, the highest amounts of PFOA were removed, at 9.7% with 20 mg/L of the injection, which was found to be negligible. The PFHxA removal was 3.33%, which means little to no removal, and the PFNA removal was 4.6% and its concentration somewhat increased with 10 mg/L of the injection and then again decreased. The other compounds, PFPeA and PFHxS, showed a removal of 5% or lower, which was negligible. Previous research result showed little to no effect of the chlorine injection on the PFCs removal regardless of the oxidant dose [9], which was similar to the results of this study.

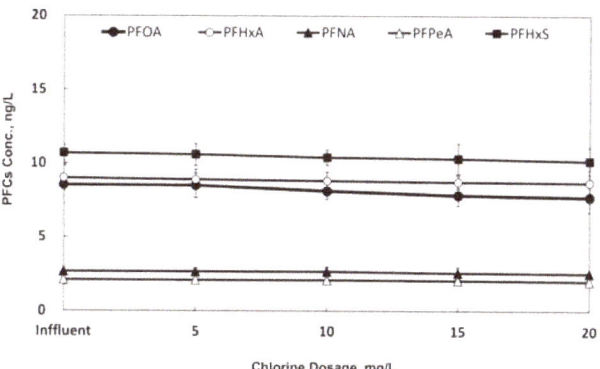

Figure 5. Variations in PFCs concentration with different chlorine dosages.

Since PFCs are stable and non-degradable, the use of a single chlorine disinfection process would not effectively remove them.

3.4.4. Removal Variation by Activated Carbon Adsorption

As shown in Figure 6, the PFNA removal at five minutes was 69.55%, which was the highest, 98.55%, at 15 min. The PFPeA removal at five minutes was 36.49%, which rose to the lowest (90.1%) at 15 min. The PFHxS removal at five minutes was 61.8%, which rose to 93.6% at 15 min. PFOA and PFHxA showed a 64.8% and 42.73% removal at 5 min of EBCT, which increased to 95.43% and 92.12% at 15 min of EBCT, respectively.

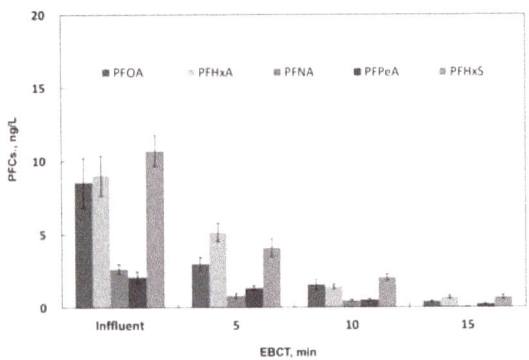

Figure 6. Variations in PFCs concentration by granular activated carbon (GAC) EBCT.

We found that 50% to 60% of PFCs was removed by five minutes of EBCT, and over 90% was removed after 15 min. PFNA showed the highest removal at 98.5%, and PFPeA showed the lowest at 90.1%. Previous research reported that the longer the length of the carbon ring in PFCs, the greater the increase in the adsorption force to the activated carbon. When the lengths of the carbon ring are identical, the adsorption force of the activated carbon to sulfone substituent would be stronger than that to carboxyl substituent [14]. In this study, the removal of PFNA, which has a large number of carbons, was highest. For PFHxA and PFHxS in which the number of carbons was identical, more of PFHxS, which has a sulfone substituent, would be removed.

The activated carbon adsorption test showed relatively higher removal, and since the surface of the activated carbon is hydrophobic, it is useful for removing hydrophobic PFCs.

3.5. Proposal of an Optimal PFCs Treatment Process

Based on the analysis results of each unit process and the pilot test effluent, we reviewed an efficient treatment process for removing PFCs and list the PFCs process efficiencies and cumulative removal efficiencies in Table 3. PFCs are hard to degrade and remove through the existing biological treatment methods and are thus considered non-degradable [9].

Table 3. Summary of PFCs applicability by various treatments.

Process		PFCs									
		PFOA		PFHxA		PFNA		PFPeA		PFHxS	
		Bat *	Conti **	Bat	Conti	Bat	Conti	Bat	Conti	Bat	Conti
Bioreactor		X	X	△	△	X	X	△	△	X	X
Coagulation-Sedimentation		X	△	X	△	△	△	△	△	△	△
Ozonation		X	△	X	△	X	△	X	△	X	△
Activated carbon treatment	EBCT 5 min	◎	◎	O	O	◎	◎	O	O	O	◎
	EBCT 15 min	◎	◎	◎	◎	◎	◎	◎	◎	◎	◎
Chlorination		X	◎	X	◎	X	◎	X	◎	X	◎

Note: X: removal efficiency <10%, △: removal efficiency 10–40%, O: removal efficiency 40–70%, ◎: removal efficiency >70%; *: Batch process; **: Continuous process.

For the activated carbon adsorption, the longer the length of the carbon ring of the PFCs, the more the adsorption force of the activated carbon increases. If the lengths of the carbon rings are identical, the adsorption force of the activated carbon is stronger with the sulfone substituent than with the carboxyl substituent [14] As for the adsorption and removal of PFCs using zeolite, the higher the Si content in zeolite, the more the PFCs adsorption capacity increases. However, this treatment is reported to have a smaller adsorption capacity than treatment with activated carbon [15]

Activated carbon adsorption showed a relatively higher PFCs removal than the other treatments. Therefore, it would be effective for PFCs removal in the wastewater treatment process. The PFHxS removal in the single ozone process or the single activated carbon process was low, but improved with the continuous ozone treatment process followed by the activated carbon process. Accordingly, the implementation of an oxidization process, such as the ozone process followed by an adsorption process using activated carbon, for example, is considered to be the most desirable process with a higher chance of removal.

4. Conclusions

We surveyed the PFCs pollution sources, targeting the influent and effluent of wastewater treatment plants, and conducted a pilot test with influent to review whether PFCs could be removed using physiochemical methods. This study produced the following results.

(1) The influent/effluent removal of the existing biological treatment process was able to remove some PFNA, PFPeA, and PFHxS, but most of the other compounds could not be removed through biological treatment.

(2) The correlational analysis results showed that there was a high correlation in the wastewater among PFOA, PFHxA, and PFNA, which contain carboxyl groups. The PFCs correlational research did not produce data that could be representative. Thus, further research is required.

(3) The pilot batch test with the influent to the treatment plant showed that about 10% of PFCs was removed by an injection of 40 to 50 mg/L of Alum in the coagulation-sedimentation test. For the pilot continuous test, the removal in the ozone test and chlorine injection test was negligible regardless of the oxidant dose. The activated carbon adsorption test showed that about 60% to 70% of PFCs was removed after EBCT 5 min, and removing over 90% of PFCs after 15 min EBCT was found to be possible.

(4) The pilot test results showed that PFHxS removal in single ozone or activated carbon process was low. However, it improved in the continuous ozone process followed by the activated carbon process. This demonstrated that ozone oxidization would promote the removal during the next process. Therefore, the continuous oxidization process, such as the ozone process, followed by an adsorption process using activated carbon, would be a desirable process enabling high removal.

Appl. Sci. **2019**, *9*, 2500

Author Contributions: Conceptualization, S.-h.L.; Methodology, S.-h.L.; Validation, S.-h.L. and Y.-j.C.; Formal analysis, Y.-j.C.; Investigation, S.-h.L.; Writing—original draft preparation, B.-D.L.; Writing—review and editing, B.-D.L.; Visualization, M.L.; Supervision, S.-h.L.; Project administration, S.-h.L.; Funding acquisition, M.L.

Funding: This work was supported by the Korea Environment Industry & Technology Institute (KEITI) through Public Technology Program based on Environmental Policy, funded by Korea Ministry of Environment (MOE) (2016000200008).

Conflicts of Interest: The authors declare no conflict of interest.

References

1. Cho, C.R.; Eom, I.C.; Kim, E.J.; Kim, S.J.; Choi, K.H.; Cho, H.S.; Yoon, J.H. Evaluation of the level of PFOS and PFOA in environmental media from industrial area and four major river basin. *J. Kor. Soc. Environ. Anal.* **2009**, *12*, 296–306.

2. Cho, C.; Cho, J.-G.; Eom, I.-C.; Lee, B.-C.; Kim, S.-J.; Choi, K.; Yoon, J. Bioconcentration of perfluorinated compounds in fish from Gulpo stream. *Environ. Health Toxicol.* **2010**, *25*, 229–240.

3. Cho, C.R.; Lee, D.H.; Lee, B.C.; Kim, S.J.; Choi, K.H.; Yoon, J.H. Residual concentrations of perfluorinated compounds in water samples of Anseong and Gyeongan streams and their spectroscopic characteristics. *J. Kor. Soc. Environ. Anal.* **2010**, *13*, 226–236.

4. Hardell, E.; Kärrman, A.; van Bavel, B.; Bao, J.; Carlberg, M.; Hardell, L. Case–control study on perfluorinated alkyl acids (PFAAs) and the risk of prostate cancer. *Environ. Int.* **2014**, *63*, 35–39.

5. *United Nations Environmental Programme*; UNEP/POPS/COP.4/38; United Nations: New York, NY, USA, 2009.

6. Ding, G.H.; Frmel, T.; van den Brandhof, E.J.; Baerselman, R.; Peijnenburg, W.J. Acute toxicity of poly- and perfluorinated compounds to two cladocerans, *Daphnia magna* and *Chydorus sphaericus*. *Environ. Toxicol. Chem.* **2012**, *31*, 605–610. [CrossRef] [PubMed]

7. Wang, Y.; Niu, J.; Zhang, L.; Shi, J. Toxicity assessment of perfluorinated carboxylic acids (PFCAs) towards the rotifer *Brachionus calyciflorus*. *Sci. Total Environ.* **2014**, *491–492*, 266–270. [CrossRef] [PubMed]

8. Stahl, L.L.; Snyder, B.D.; Olsen, A.R.; Kincaid, T.M.; Wathen, J.B.; McCarty, H.B. Perfluorinated compounds in fish from U.S. urban rivers and the Great Lakes. *Sci. Total Environ.* **2014**, *499*, 185–195. [CrossRef] [PubMed]

9. Park, C.-G. Study on concentration distribution and removal characteristics of micro-pollutants in the middle Nakdong river basin. Ph.D. Thesis, Yeungnam University, Gyeongsan, Korea, 2013.

10. Roth, N.; Wilks, M.F. Neurodevelopmental and neurobehavioural effects of polybrominated and perfluorinated chemicals: A systematic review of the epidemiological literature using a quality assessment scheme. *Toxicol. Lett.* **2014**, *230*, 271–281. [CrossRef] [PubMed]

11. Shin, M.Y.; Im, J.K.; Kho, Y.L.; Choi, K.S.; Zoh, K.D. Quantitative determination of PFOA and PFOS in the effluent of sewage treatment plants and in Han river. *J. Environ. Health Sci.* **2009**, *35*, 334–342. [CrossRef]

12. Ahrens, L.; Barber, J.L.; Xie, Z.; Ebinshaus, R. Longitudinal and latitudinal distribution of perfluoroalkaly compounds in surface water of the Atlantic ocean. *Environ. Sci. Technol.* **2009**, *43*, 3122–3127. [CrossRef] [PubMed]

13. So, M.K.; Miyake, Y. Perfluorinated compounds in the Pearl river and Yangtze river of China. *Chemosphere* **2007**, *68*, 2085–2095. [CrossRef] [PubMed]

14. Du, Z.; Deng, S.; Bei, Y.; Huang, Q.; Wang, B.; Huang, J.; Yu, G. Adsorption behavior and mechanism of perfluorinated compounds on various adsorbents-a review. *J. Hazard Mater.* **2014**, *274*, 443–454. [CrossRef] [PubMed]

15. Ochoa-Herrera, V.; Sierra-Alvarez, R. Removal of perfluorinated surfactants by sorption onto granular activated carbon, Zeolite and sludge. *Chemosphere* **2008**, *72*, 1588–1593. [CrossRef] [PubMed]

Article

Ultrafiltration/Granulated Active Carbon-Biofilter: Efficient Removal of a Broad Range of Micropollutants

Christian Baresel *, Mila Harding and Johan Fang

IVL Swedish Environmental Research Institute, Box 210 60, 100 31 Stockholm, Sweden;
mila.harding@ivl.se (M.H.); johan.fang@ivl.se (J.F.)
* Correspondence: christian.baresel@ivl.se; Tel.: +46-010-7886606

Published: 18 February 2019

Abstract: Pharmaceutical residues, and other organic micropollutants that pass naturally through the human body into sewage, are in many cases unaffected by treatment processes at conventional wastewater treatment plants (WWTPs). Accumulated in the environment, however, they can significantly affect aquatic ecosystems. The present study provides an evaluation of a treatment system for the removal of pharmaceutical residues and other micropollutants. The system is based on a Membrane Bioreactor (MBR), including ultrafiltration (UF), followed by a biofilter using granulated active carbon (GAC) as filter material. It was found that all investigated micropollutants, such as pharmaceutical residues, phenolic compounds, bacteria and microplastic particles, present in wastewater, could be removed by the treatment system to below detection limits or very low concentrations. This shows that the combination of filtration, adsorption and biodegradation provides a broad and efficient removal of micropollutants and effects. The tested treatment configuration appears to be one of the most sustainable solutions that meets today's and future municipal sewage treatment requirements. The treatment system delivers higher resource utilization and security than other advanced treatment systems including solely GAC-filters without biology.

Keywords: water quality; Membrane Bioreactor; GAC-biofilter; sewage treatment; micropollutants; pharmaceutical residues; activated carbon

---

1. Introduction

Micropollutants (MPs), generally summarizing pharmaceutical residues and other emerging substances, pass through traditional wastewater treatment plants (WWTPs) and end up in the receiving waters and sludge. Various studies reported recipient concentrations with expected effects on aquatic organisms [1–6]. MPs released via WWTPs may also enter the aquatic food web and cause effects in higher organisms such as fish-eating birds or mammals including humans. Therefore, an increasing demand for supplementary treatment at today's WWTPs for the efficient removal of micropollutants has become obvious.

As current WWTPs are usually unable to remove micropollutants, a number of various treatment technologies have been proposed and evaluated through several large projects, such as those in References [7–9] and the Swedish MistraPharma. Technologies tested include, among others, membrane separation (reverse osmosis, nanofiltration, ultrafiltration), advanced oxidation processes (ozonation, UV-light in combinations with hydrogen peroxide and titanium dioxide) and activated carbon (powdered and granular activated carbon). Especially in Germany and Switzerland, advanced treatment technologies have been tested on a large scale [10,11]. Also in Sweden, technologies have been tested [12–17]. In Germany, Austria and Sweden, first full-scale installations have already been accomplished.

Most of available studies on treatment technologies focused on the efficiency of removing pharmaceutical residues. The processes that are most often considered effective are the treatment with ozone or activated carbon. Generally, ozone treatment implies both a direct chemical reaction of the ozone molecule as well as indirect reactions with hydroxyl radicals, breaking specific chemical bonds within the targeted substances. There exist several studies investigating complementary treatment by ozone [11–15,17–28]. Results indicate that while ozone oxidation generally provides a removal effect on many targeted substances, a sufficient removal of some substances may not be achieved even at very high ozone doses. Further, the main disadvantage of ozone treatment is the fact that the process does not completely degrade most substances. These may be transformed into other substances, normally without aromatic structures. Some of these metabolites might be more or less toxic and require thus an extra treatment step after ozonation [19–22,25–28]. These problems can be handled using a more integrated treatment setup as proposed by [14] using an ozonation step between bio-sedimentation and post-denitrification processes. This configuration is realized as Sweden's first full-scale installation of micropollutant removal at municipal WWTPs. Other challenges when using ozonation are an additional high-energy demand and working environment issues at WWTPs.

The use of activated carbon (AC), either as powdered activated carbon (PAC) or granulated activated carbon (GAC), has been investigated in numerous studies [11,16,29–35]. The use of AC is a widespread technology to remove various pollutants from water. Especially in treatment of fresh water for drinking water production, technical systems using either PAC or GAC have been applied for many years. Thus, significant knowledge on setup and operation of such systems is available. The main advantage of using activated carbon is a broad and effective removal of MPs and that no by-products are generated. During regeneration or destruction of activated carbon, the adsorbed pollutants are destroyed. The currently high environmental impact of AC-applications is caused by the immense energy and resource utilization during production and regeneration of activated carbon. This is identified as a drawback of the technology that can only be solved by increasing the AC-capacity or utilization in different ways or using biochar based on organic waste such as sewage sludge [36].

Filter systems based on GAC are common and a potential biological activity inside the filter will affect adsorbed organic compounds and the overall filter performance. Biological activity, however, is, next to suspended solids in the inflow, the main reason for clogging problems representing a key operational challenge for GAC-filter systems. An excellent pre-treatment and particle-free process waters are generally easier to handle and could improve the application potential of GAC-biofilter systems.

Membrane Bioreactor (MBR) systems are currently considered at more and more WWTPs to meet challenges with increased load as well as more stringent effluent quality requirements. Several WWTPs in Sweden, such as the Stockholm Water and Waste Company (Stockholm Vatten och Avfall), Sweden's largest water service organization, are replacing their existing conventional activated sludge process (CAS) with an MBR. After upgrading, the new process will be one of the world's largest MBR facility with a capacity of 1.6 million PE (predicted load year 2040). MBRs combine the biological activated sludge process with membrane separation, which provide distinct advantages over the CAS. Advantages include a significant better effluent (permeate) quality regarding particles, disinfection capabilities due to the membrane pore size, higher volumetric loading due to higher sludge concentrations in the biology, reduced footprint and process flexibility towards influent changes. Even the treatment of micropollutants (MPs) may be more efficient using MBRs compared to traditional treatment systems. This is partly explained by the fact that MP attached to particles can efficiently be removed by filtration, which also includes for example microplastics.

MBRs have been used for a number of decades but first in the last decade, MBRs gained more attention for the treatment of both municipal and industrial wastewater. This is much due to a significant cost reduction of membranes and process development decreasing energy requirements [37–41], which also implies a significant increase in current and planned installations worldwide.

The aim of this research work was to investigate and evaluate the long-term removal efficiency of a number of micropollutants including pharmaceutical residues, microplastics and so forth, by using a GAC-biofilter applied to MBR-effluent. Through actual pilot process setup as designed for a real WWTP, the current work evaluates the performance of the studied system and its potential role in the way forward for micropollutants removal.

While the removal efficiency of activated carbon and ultrafiltration has been evaluated in previous studies as described, no long-term evaluation of the combination of an MBR process and a GAC-biofilter has been done. The current work further is novel as it focusses on the pharmaceutical residue removal by the biological activity in the GAC-biofilter. Previous evaluations of GAC-filtration systems focus on the adsorption capacity of the activated carbon. The combination with an MBR-process is motivated by recent developments in the sewage treatment that will result in a significant increase in MBR-installations worldwide and thus the relevance of the current study.

2. Materials and Methods

2.1. Pilot Characteristics

For the evaluation in long-term tests, an MBR-pilot was applied as the main treatment process. IVL Swedish Environmental Research Institute and the Stockholm Water and Waste Company have together set up, and since September 2013 operated, a pilot-scale treatment line with a capacity corresponding to 0.015% of the total Henriksdal WWTP facility (design year 2040). The pilot is located at the Research and Development (R&D) facility, Hammarby Sjöstadsverk (www.hammarbysjostadsverk.se). Wastewater treated by the pilot is taken from the untreated inflow to Stockholm's main WWTP Henriksdal and filtered through a 3 mm strainer (Figure 1). The flow into the pilot is proportional to the flow to the main WWTP and the hydraulic retention time (HRT) in the biological reactor corresponds to 10 hours at average flow. The pilot consists of a primary clarifier, a biological reactor with a total volume of about 29 m^3 including anoxic and aerobic zones, followed by an ultra-filtration (UF). Anoxic and aerobic zones account each for 50% of the process volume. The membrane tank had a total volume of 13 m^3. Nitrate is recirculated from the beginning of the post-denitrification zone to the beginning of the pre-denitrification zone and sludge is recirculated from the UF to the beginning of the pre-denitrification zone. The ultrafiltration consists of two modules with Flat Sheet membrane type MFM 100 from Alfa Laval (Denmark). The UF units are operated intermittently with relaxation times of 2 minutes after 10 minutes of operation. The nominal pore size is 0.2 microns with a minimum and maximum pore size of 0.17 microns and 0.26 microns, respectively. The total membrane area per module is 79.64 m^2 spread over 44 membrane sheets. A more detailed description of the MBR-pilot configuration and operational characteristics is provided by Baresel et al. (2017) [42].

The nominal pore size of the UF in the MBR-system of 0.2 microns implies that particles of larger size are efficiently removed from the wastewater, including microplastics, bacteria and pathogens. The evaluation of the MBR-process shows that targeted effluent qualities of <0.2 mg TP/L and 6 mg TN/L are achieved under various loads. Previous analyses of pharmaceutical residues in other MBR-effluent showed no increased removal effect of pharmaceuticals by the MBR-process compared to the CAS-process [43–45].

Particle-free MBR treated wastewater was pumped at a constant flow of 1400 L/h to the pilot GAC-filter (Figure 1) with a surface area of 0.3 m^2. The filter consists of a 10 cm thick sand bed on the bottom and a 1 m layer of commercial granulated carbon (Filtrasorb 400, Chemviron Carbon, density ~ 0.5 kg/L). On the filter bottom, there are a number of nozzles for backwash from an equalization tank equipped with a continuous measurement of the suspended matter content to the top of the GAC filter. The GAC-filter was originally constructed as sand filter and operated as such for many years before it was used as GAC-biofilter in this study.

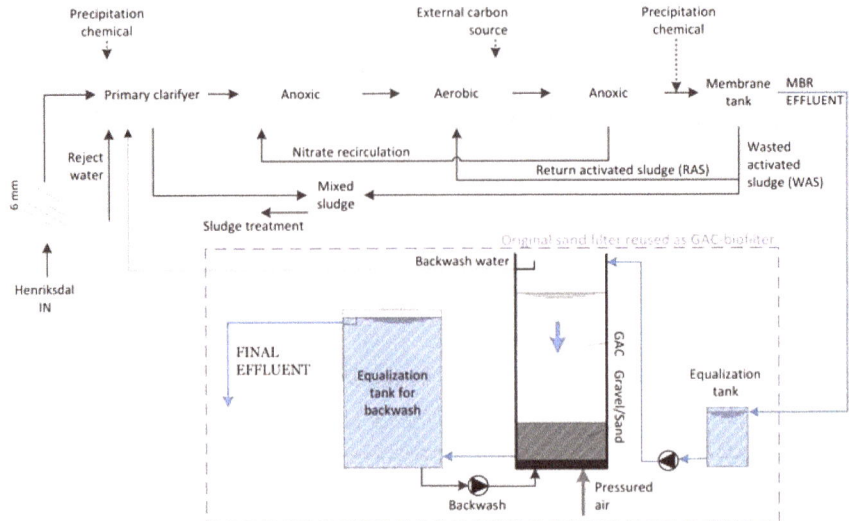

Figure 1. Schematic illustration of the pilot setup including a membrane bioreactor MBR and granulated active carbon (GAC)-biofilter for removal of micropollutants.

With the indicated normal flow and filter volume, a contact time in the filter (HRT or EBCT (Empty Bed Contact Time)) of 13 min was maintained. These operational parameters were based on related pilot trials [46,47] where different residence times in a GAC-filter were tested with water treated in a temporary MBR-pilot.

The water passed through the filter and was collected in an equalization tank for backwash. Backwash consisted of a sequence of pulses of pressured air to terminate eventual pressed layers and backwash with water from the equalization tank. Backwash water was diverted back to the main inflow of the MBR-pilot. The filter was open and the driving force through the column was the difference in level between the water in the column and the level of the outlet. The water level in the GAC-filter was regulated via level gauge controlling the valve opening for outgoing water. The level of control was 40–50 cm above the filter bed.

2.2. Sampling and Analysing

The long-term test lasted for almost two years. Automatic samplers continuously collected flow-propositional samples of the untreated wastewater, the MBR-effluent and the final effluent after the GAC-biofilter and stored them cooled. Each week, composite samples were collected and frozen. During start-up and after some weeks of operation, weekly composite samples were sent for analyses. With the analyses at hand, coming weeks for analyses were planned for or previously collected samples were added in order to cover periods with significant changes. Grab samples for bacteria analyses were collected at the final day of a sampling campaign.

Investigated micropollutants include a wide range of relevant pharmaceuticals and other emerging substances, oestrogen effect, bacteria and microplastics (see Table S1 for details about investigated substances). Generally, triplicate analyses were performed on all samples. Only certified laboratories were utilized in the project. Thus, standard analytical methods for all analyses and are not described in detail here. Pharmaceuticals and microplastic particles were analysed at IVL's own certified laboratory using the following methods.

Pharmaceuticals were analysed using aliquots of 100 to 200 mL thawed composite samples that were spiked with 50 µL internal standard carbamazepine-13C15N (2000 ng/mL) and ibuprofen-D3 (2000 ng/mL). One millilitre of 0.1 wt% ethylenediaminetetraacetate (EDTA-Na2) dissolved in methanol:water (1:1) was added. Prior to extraction using solid phase extraction (SPE) cartridges (Oasis HLB, 6 mL, Waters), the sample was shaken. Cartridges were conditioned with methanol followed by Milli-Q (MQ) water. Thereafter, the samples were applied to the columns at a flow rate of two drops per second. The substances were eluted from the SPE cartridges using 5 mL methanol followed by 5 mL acetone. The supernatants were transferred to vials for final analysis on a binary liquid chromatography (UFLC) system with auto injection (Shimadzu, Japan). The chromatographic separation was carried out using gradient elution on a C18 reversed phase column (dimensions 50×3 mm, 2.5-µm particle size, XBridge, Waters, UK) at a temperature of 35 °C and a flow rate of 0.3 mL/ min. The mobile phase consists of 10 mM acetic acid in water.

In addition to pharmaceuticals and phenolic compounds in water, also the contents in the filter material was analysed at the end of the experiment. Pharmaceuticals residues in the carbon were determined after representative samples were taken, dewatered and freeze-dried. The substances were extracted with acetone: acetic acid (20:1). The eluate was then treated as for the water samples.

Even so, replicate analyses have been performed, complex wastewater and filter material matrixes imply challenges during sample preparation and analyses. For example, other organic substances can reduce the recovery during sample preparation and affect the signal during analysis or some substances to be analysed can interact with free ions from the matrix and form chelate complex, which result in reduced recovery and detection. As the test are based on real wastewater including the daily, weekly and seasonal quality variation, analyses uncertainty varies during the 2 years of analyses as also the water matrix varies. Therefore, only average values of replicate analyses are presented in this study.

Microplastic particles were analysed by following method [48] commonly used in screenings in Nordic countries as standards for microplastic analyses are not yet established. The water samples were filtered through filters with a mesh size of 20 µm and the material collected on the filters was analysed with a stereo microscope (50 times magnification). All microplastic particles were counted and divided into three groups according to their shape—plastic fragments, plastic flakes and plastic fibres. The term plastic flake was used for very thin particles, whereas thicker particles were called plastic fragments. The term microplastics or plastic particles refer to all three groups. In addition to the microplastics, also non-synthetic fibres of anthropogenic origin were counted. This included textile fibres of for example cotton but not cellulose from toilet paper.

Material suspended before the GAC-filter was measured with an online meter of the Züllig COSMOS 25. The total suspended solids (TSS) was determined by standard method (SS 02 81 12-3) and BOD5 with WTW Oxitop. pH was determined with a hand meter (pH 3110 from WTW) and colour, transmission and absorbance at 254 nm were determined using a spectrophotometer—WTW photoLAB 6600. TOC was determined according to standard method (SS-EN 1484).

3. Results and Discussions

Many pharmaceutical substances were already removed in the MBR to levels below the reporting limit. Compared with the full-scale WWTP Henriksdal, the MBR resulted in lower concentrations of for example furosemide, bisoprolol, metoprolol and sertraline. Reasons for the better treatment may be the higher sludge age providing an enhanced biodegradation or certain adsorption to the membrane (sertraline) as also reported by [49].

For the evaluation of the removal efficiency in the GAC-biofilter, only compounds that were always quantifiable before the filter and at least once quantifiable after the filter were considered in the assessment. Figure 2 shows filter effluent concentrations as a percentage of the influent concentration for these compounds. Values below LOQ (limit of quantification) are here set as LOQ/2. The x-axis is graded with the number of Empty Bed Volumes (EBVs) that passed the GAC-biofilter. 60 000 EBV corresponds to 120 m^3 water/kg GAC in the filter, that is 574 days after the start of the experiment when the tests had to be stopped due to reconstruction of the MBR-pilot. The figure also shows the targeted removal efficiency by the project and common design criteria for GAC-filter where only adsorption of micropollutants is considered.

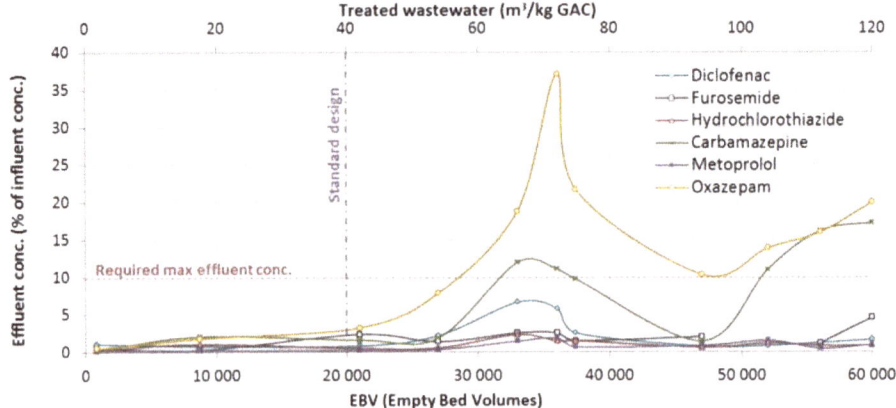

Figure 2. Average reduction of various pharmaceuticals during the whole project period.

Figure 2 shows a good removal of all substances immediately after the operation of the filter was initiated. This is mainly explained by a high adsorption capacity of the fresh GAC as also reported by [16,32]. Up to about 25,000 EBV (about 50 m^3/kg GAC), the removal was very good for all compounds. Then elevated effluent concentrations for some compounds can be observed. The level of oxazepam and carbamazepine increased to more than 10% of incoming concentrations and the diclofenac concentration exceeded 5 % of the incoming concentration. The GAC-biofilter operation, however, was continued to evaluate the long-term removal efficiency of the GAC-biofilter for all substances. As the figure indicates, the removal efficiency improved again without any changes in operating mode. A similar recovery was also noted in earlier trials with a GAC-filter treating the effluent of a WWTP operated as conventional activated sludge process [16]. However, analysed concentrations do not indicate any significant changes in the incoming load to the GAC-biofilter and can thus not explain the temporal increase of concentrations in the GAC-biofilter effluent. In addition, no other test parameters, for example water temperature, changed in a way that could explain observed removal variations in Figure 2. A bio-regeneration inside the GAC-biofilter may be a potential explanation. Bio-regeneration was reviewed before as a more sustainable alternative to conventional regeneration methods but focus has been on bio-regeneration in offline filters [50]. However, it is mentioned that the same process may also take place in GAC-filters with an established biological activity. Although there have been some good research efforts in bio-regeneration, there is still not much known about the factors affecting the regeneration process [50].

After about 50,000 EBV (about 100 m^3/kg GAC), another increase of effluent concentrations can be observed for oxazepam and carbamazepine. The increase of furosemide is not certain as the level was below LOQ and the high percentage is due to very low concertation levels. At 60,000 EBV (about 120 m^3/kg GAC), the experiment was terminated due to a required modification of the MBR-pilot. The sampling frequency with weekly composites samples does not provide any information about actual load variations with a higher resolution. It is not self-evident whether the removal in the GAC-biofilter is defined as a certain percentage of incoming concentration, a certain amount per unit of time or down to a certain residual content. Probably it is a combination as also suggest by [51]. Evaluating the actual concentrations in the effluent show that the corresponding curves to Figure 2 have roughly the same shape. However, for Oxazepam the peak appears somewhat later. This is probably due to varying concentrations in the inflow to the filter. Table S2 in the Supplementary Materials shows analysed concentrations of all compounds before and after the GAC-biofilter at the end of the test.

At the end of the experiment, samples of the used GAC were taken from various levels of the filter bed. Higher concentrations of pharmaceutical residues were observed in samples from the top layer of the filter bed despite frequent backwashing during the 2 years of operation. This implies that the frequent backwash did not affect the concertation profile in the filter as also suggested by [16]. A screening analysis of the used GAC indicated very small changes in the size distribution of the carbon particles, which indicates a very morphologically stable carbon. Despite this, a decomposition of carbon particles could be observed in a larger proportion of small particles, which also implies that the number of larger particles was reduced.

After the pilot operation was finalized, a rough mass-balance for the removed pharmaceutical substances by the GAC-biofilter was established. Amounts of substances removed were calculated based on all analyses and the assumption that the concentrations varied linearly between analyses events. For the GAC-biofilter effluent, concentrations below LOQ have been considered as LOQ/2. Analysis of the backwash water showed high levels of suspended biological material, 200-300 mg/L but very low levels of pharmaceutical substances. The recirculation of these substances back to the inflow of the MBR-pilot via backwash water was estimated to be less than 1% of the total load to the GAC-biofilter. The difference between incoming and outgoing mass flows is then determined as the amount removed by the GAC-biofilter. As most of the compounds are considered as stable, they should then be contained in the used GAC. Table 1 shows the amounts of considered pharmaceutical residues found in the GAC. Presented results are corrected for the exchange of the respective substance in the extraction of new spiked carbon. The yield was between 50 and 100 % for the various compounds.

Table 1. Total amount of removed pharmaceutical residues in the GAC-biofilter.

	Total Removed	Analysed in GAC	Adsorbed
	mg/kg GAC	mg/kg GAC	%
Citalopram	29.2	1.09	3.7
Diclofenac	67.9	0.13	0.2
Furosemide	49.2	0.57	1.2
Hydrochlorothiazide	143.4	3.97	2.8
Ibuprofen	8.1	0.01	0.1
Carbamazepine	41.2	13.1	31.8
Metoprolol	82.5	3.15	3.8
Oxazepam	54.3	7.03	12.9
Propranolol	6.7	0.87	12.0

Despite several uncertainties in the mass balance, it is clear that most of the considered compounds that efficiently have ben remove from the wastewater, were not found in the analysed GAC. After initially being adsorbed by the GAC, the compounds may have been broken down biologically by the established biological activity in the filter as also discussed by [50]. Even a metabolization of the substances in the filter may be possible, as applied analytical methods could not measure metabolites of the considered pharmaceuticals. Measured COD/TOC in the effluent of the filter was stable and the change of Spectral Absorption Coefficient (SAC) over the filter remained stable. A microbial screening of the biofilm in the filter material and a related quantification of the removal of substances by either adsorption or biodegradation is difficult [50] and they were not performed in this project. The assessment of the adsorbed contaminants in the GAC as sown in Table 1, supports the assumption of a bio-regeneration in the GAC-biofilter, which extends the lifetime of the system (see also [50]).

In general, the MBR-process provides a high quality, particle-free effluent compared to traditional activated sludge processes. Bacteria, including multiresistant bacteria, of all sizes larger than the membrane pore size were efficiently removed from the wastewater by the MBR-process. However, very low concentrations (<65 cfu/100 mL) of bacteria were still detected in the MBR-effluent. It could, however, not be determined if these bacteria originated from sample contamination or contact of the permeate with the atmosphere. Both aspects are almost impossible to avoid in sewage treatment environments. Total coliforms in the treated MBR-effluent were further reduced with >85% by the

GAC-biofilter. Interestingly, faecal coliform removal was absent during the first weeks of operation while a reduction of more than 90% was achieved after 3 months of operation. This might be explained by the established biology in the filter that outcompetes faecal coliforms.

The studied phenolic compounds triclosan and bisphenol A were reduced to below detection limit. Most of nonylphenol and octylphenol was removed as well. It was difficult to extract several of the phenolic compounds from the carbon. Thus, a mass balances for these compounds over the GAC-biofilter was not possible to perform. High levels of nonylphenol, triclosan and bisphenol A in the backwash water suggested that they were largely bound to the flushed biomass and thus returned to the biological process in the main treatment.

Not a single microplastic particle was detected in the MBR-effluent (removal efficiency 100%), whereas effluent water from the full-scale CAS-process including a final sand filtration contained both plastic fibres and plastic fragments (removal efficiency 90.7 %). Non-synthetic fibres were found in both MBR and CAS effluents.

Compared to complimentary treatment of the final effluent from the main WWTP Henriksdal, that is the same influent water, [16,52], a significant reduction of clogging and backwash frequency was achieved in the GAC-biofilter when treating MBR-effluent. Both aspects have a direct impact on the operational cost of the GAC-biofilter system. The better quality of MBR-effluent compared to traditional CAS-effluent (even with sand filtration [15]) provided better conditions for the GAC-biofilter operation.

The initial adsorption of pharmaceuticals substances and thus concentration build-up on the filter material provides good conditions for the establishment of a biology. Compared to treatments system that use the same technology combination but in a different order [51], are more specialized biology may be able to establish as easier degradable organic contaminants have already been removed by the preceding MBR-process. High oxygen concentrations in the effluent from the MBR due to continuous air scouring of the membranes may enhance the biological breakdown of organic micropollutants in the following biofilter.

The evaluation of the pilot operation and related removal of micropollutants indicates that some aspects need further confirming experiments in order to utilize the findings in the most optimal way. Here the recovery of the removal capacity in the biofilter is the most interesting aspect for further investigation as also pointed out by [50].

4. Conclusions

The combination of Membrane Bioreactor (MBR) and biofilter with granulated activated carbon (GAC) as filter material have not received the same attention as resource efficient removal alternative for micropollutants as other technologies. The combination of an enhanced biology and ultrafiltration in the MBR, followed by adsorption and biological degradation in the GAC-biofilter, however, is a powerful treatment alternative. Considering the increasing number of MBR installations in municipal sewage treatment worldwide, the treatment combination has a significant potential to meet requirements for less micropollutant discharge to the environment in a resource-efficient way, especially in large WWTPs. The combination of an MBR system and GAC-biofilter cannot only remove a broader range of micropollutants than ozonation. Further, the system does not impose any risk of the formation of toxic residues and has greater improvement potential regarding environmental sustainability and costs.

The long-term evaluation of the GAC-biofilter subsequent of an MBR-treatment shows that about 90–98 % of the pharmaceutical residues could be removed from the water. The assessment illustrates the importance of long-term tests to determine the actual capacity of a biological active filter. This combination of the different treatment technologies and associated removal processes not only facilitates a more efficient removal of pharmaceutical residues, but it also prolongs the lifetime of the filter material. The performed analyses and mass balances show that only a minor amount of the removed substances was adsorbed to the filter material. The majority of the removed pharmaceutical substances was broken down by the established biology in the filter. The results further indicate that the biological activity in the GAC-biofilter can provide a bio-regeneration of the GAC by decomposing targeted substances and thereby restoring adsorption capacity of the GAC. This finding may have significant impact on the overall resource efficiency of such treatment systems, both regarding costs and overall environmental impact of the additional treatment.

The investigation further shows the advantage of using GAC as filter material as the high adsorption capacity of GAC ensures a high removal efficiency right from the start-up of the filter even so a microbial community on the GAC surface requires time to establish. The initial adsorption of pharmaceuticals substances and thus concentration build-up on the filter material provides good conditions for the establishment of a specialized biology.

In general, the project results show that the combination of an MBR-process with a GAC-biofilter provides a complementary treatment system able to meet various demands for efficient sewage treatment to low effluent concentrations of organics, nutrients, suspended solids and micropollutants. As requirements on sewage treatments including the removal of micropollutants will continuously become stricter, the investigated treatment system of MBR and GAC-biofilter may be one of the most attractive solution for a resource-efficient removal of a broad range of micropollutants from sewage. Even so, more research on this treatment system is necessary; the current study clearly indicates the potential of the system.

Supplementary Materials: The following are available online at http://www.mdpi.com/2076-3417/9/4/710/s1.

Author Contributions: The three authors have made equal substantial contributions to the planning, setup and operation of the pilot-test and interpretation of the results.

Funding: This research received no external funding.

Conflicts of Interest: The authors declare no conflict of interest.

References

1. Brodin, T.; Fick, J.; Jonsson, M.; Klaminder, J. Dilute Concentrations of a Psychiatric Drug Alter Behavior of Fish from Natural Populations. *Science* **2013**, *339*, 814–815. [CrossRef] [PubMed]

2. Deblonde, T.; Cossu-Leguille, C.; Hartemann, P. Emerging pollutants in wastewater: A review of the literature. *Int. J. Hyg. Environ. Health* **2011**, *214*, 442–448. [CrossRef] [PubMed]

3. Fick, J.; Lindberg, R.H.; Kaj, L.; Brorström-Lundén, E. *Results from the Swedish National Screening Programme 2010, Subreport 3, Pharmaceuticals*; B2014; IVL Swedish Environmental Research Institute: Stockholm, Sweden, 2011.

4. Fick, J.; Lindberg, R.H.; Schwesig, D.; Gawlik, B.M. EU-wide monitoring survey on emerging polar organic contaminants in wastewater treatment plant effluents. *Water Res.* **2013**, *47*, 6475–6487. [CrossRef]

5. Kim, S.D.; Cho, J.; Kim, I.S.; Vanderford, B.J.; Snyder, S.A. Occurrence and removal of pharmaceuticals and endocrine disruptors in South Korean surface, drinking, and waste waters. *Water Res.* **2007**, *41*. [CrossRef] [PubMed]

6. Vasquez, M.I.; Lambrianides, A.; Schneider, M.; Kümmerer, K.; Fatta-Kassinos, D. Environmental side effects of pharmaceutical cocktails: What we know and what we should know. *J. Hazard. Mater.* **2014**, *279*, 169–189. [CrossRef] [PubMed]

7. Rempharmawater. Ecotoxicological Assessments and Removal Technologies for Pharmaceuticals in Wastewaters. Project Reference: EVK1-CT-2000-00048. 2003. Available online: http://www.unina.it (accessed on 22 August 2018).

8. POSEIDON. Assessment of Technologies for the Removal of Pharmaceuticals and Personal Care Products in Sewage and Drinking Water Facilities to Improve the Indirect Potable Water Reuse. Project Reference: EVK1-CT-2000-00047. 2004. Available online: http://www.eu-poseidon.com (accessed on 22 August 2018).

9. RiSKWa. *Risk Management of Emerging Compounds and Pathogens in the Water Cycle*; Bundesministerium für Bildung und Forschung (BMBF): Bonn, Germany, 2013; Available online: http://www.bmbf.riskwa.de (accessed on 22 August 2018).

10. Abegglen, C.; Siegrist, H. *Mikroverunreinigungen aus Kommunalem Abwasser. Verfahren zur Weitergehenden Elimination auf Kläranlagen (Micropollutants from Municipal Wastewater. Method for Further Elimination on Sewage Treatment Plants)*; Umwelt-Wissen Nr. 1214: 210; Bundesamt für Umwelt: Bern, Germany, 2012.

11. Arge Spurenstoffe NRW. *Teilprojekt 6. Elimination von Arzneimitteln und Organischen Spurenstoffen: Entwicklung von Konzeptionen und Innovativen, Kostengünstigen Reinigungsverfahren (Elimination of Pharmaceuticals and Organic Trace Substances: Development of Concepts and Innovative, Cost-Effective Cleaning Methods)–Abschlussbericht zur Phase 2*; Arge Spurenstoffe NRW: Bochum, Germany, 2013.

12. Baresel, C.; Ek, M.; Harding, M.; Bergström, R. *Treatment of Biologically Treated Wastewater with Ozone or Activated Carbon*; B2203; IVL Swedish Environmental Research Institute: Stockholm, Sweden, 2014.

13. Baresel, C.; Dahlgren, L.; Nikolic, A.; de Kerchove, A.; Almemark, M.; Ek, M.; Harding, M.; Ottosson, E.; Karlsson, J.; Yang, J. *Reuse of Treated Wastewater for Nonpotable Use (ReUse)–Final Report*; B2219; IVL Swedish Environmental Research Institute: Stockholm, Sweden, 2015.

14. Baresel, C.; Malmborg, J.; Ek, M.; Sehlén, R. Removal of pharmaceutical residues using ozonation as intermediate process step at Linköping WWTP, Sweden. *Water Sci. Technol.* **2016**, *73*, 2017–2024. [CrossRef] [PubMed]

15. Baresel, C.; Ek, M.; Ejhed, H.; Allard, A.S.; Magnér, J.; Dahlgren, L.; Westling, K.; Wahlberg, C.; Fortkamp, U.; Søhr, S. *Handbook for the Treatment of Micropollutants at Sewage Treatment Plant–Planning and Installing Treatment Techniques for Pharmaceutical residues and Other Micropollutants*; Final Report SystemLäk Project; B2288; IVL Swedish Environmental Research Institute: Stockholm, Sweden, 2017.

16. Ek, M.; Baresel, C.; Magnér, J.; Bergström, R.; Harding, M. Activated carbon for the removal of pharmaceutical residues from treated wastewater. *Water Sci. Technol.* **2014**, *69*, 2372–2380. [CrossRef] [PubMed]

17. Wahlberg, C.; Björlenius, B.; Paxéus, N. *Läkemedelsrester i Stockholms Vattenmiljö–Förekomst, Förebyggande Åtgärder och Rening av Avloppsvatten (Pharmaceutical Residues in Stockholm's Aquatic Environment–Prevention, Prevention and Treatment of Sewage)*; Stockholm Vatten AB: Stockholm, Sweden, 2010; ISBN 978-91-633-6642-0.

18. Altmann, J.; Ruhl, A.S.; Zietzschmann, F.; Jekel, M. Direct comparison of ozonation and adsorption onto powdered activated carbon for micropollutant removal in advanced wastewater treatment. *Water Res.* **2014**, *55*, 185–193. [CrossRef]

19. Abegglen, C.; Escher, B.; Hollender, J.; Siegrist, H.; von Gunten, U.; Zimmermann, S.; Häner, A.; Ort, C.; Schärer, M. Ozonung von gereinigtem Abwasser zur Elimination von organischen Spurenstoffen. Grosstechnischer Pilotversuch Regensdorf (Schweiz) (Ozonation of purified wastewater to eliminate organic trace substances. Large-scale pilot test Regensdorf (Switzerland)). *Korrespondenz Abwasser Abfall* **2010**, *57*, 155–160.

20. Gerrity, D.; Snyder, S. Review of Ozone for Water Reuse Applications: Toxicity, Regulations, and Trace Organic Contaminant Oxidation. *Ozone Sci. Eng.* **2011**, *33*, 253–266. [CrossRef]

21. Magdeburg, A.; Stalter, D.; Oehlmann, J. Whole effluent toxicity assessment at a wastewater treatment plant upgraded with a full-scale post-ozonation using aquatic key species. *Chemosphere* **2012**, *88*. [CrossRef] [PubMed]

22. Magdeburg, A.; Stalter, D.; Schlüsener, M.; Ternes, T.; Oehlmann, J. Evaluating the efficiency of advanced wastewater treatment: Target analysis of organic contaminants and (geno-)toxicity assessment tell a different story. *Water Res.* **2014**, *50*, 35–47. [CrossRef] [PubMed]

23. Maus, C.; Herbst, H.; Ante, S.; Becker, H.-P.; Glathe, W.; Bärgers, A.; Türk, J. Guidance on the interpretation and design of ozonation plants for micropollutants elimination. *Korrespondenz Abwasser Abfall* **2014**, *61*, 998–1006.

24. Reungoat, J.; Escher, B.I.; Macova, M.; Keller, J. Biofiltration of wastewater treatment plant effluent: Effective removal of pharmaceuticals and personal care products and reduction of toxicity. *Water Res.* **2011**, *45*, 2751–2762. [CrossRef] [PubMed]

25. Stalter, D.; Magdeburg, A.; Weil, M.; Knacker, T.; Oehlmann, J. Toxication or detoxication? In vivo toxicity assessment of ozonation as advanced wastewater treatment with the rainbow trout. *Water Res.* **2010**, *44*, 439–448. [CrossRef] [PubMed]

26. Stalter, D.; Magdeburg, A.; Oehlmann, J. Comparative toxicity assessment of ozone and activated carbon treated sewage effluents using an in vivo test battery. *Water Res.* **2010**, *44*, 2610–2620. [CrossRef] [PubMed]

27. Stalter, D.; Magdeburg, A.; Wagner, M.; Oehlmann, J. Ozonation and activated carbon treatment of sewage effluents: Removal of endocrine activity and cytotoxicity. *Water Res.* **2011**, *45*, 1015–1024. [CrossRef] [PubMed]

28. Wert, E.C.; Rosario-Ortiz, F.L.; Drury, D.D.; Snyder, S.A. Formation of oxidation byproducts from ozonation of wastewater. *Water Res.* **2007**, *41*, 1481–1490. [CrossRef] [PubMed]

29. Alt, K.; Mauritz, A. Projekt zur Teilstrombehandlung mit Pulveraktivkohle im Klärwerk Mannheim (Project for sidestream treatment with powdered activated carbon in the sewage treatment plant Mannheim). *Korrespondenz Abwasser Abfall* **2010**, *57*, 161.

30. Boehler, M.; Zwickenpflug, B.; Hollender, J.; Ternes, T.; Joss, A.; Siegrist, H. Removal of micropollutants in municipal wastewater treatment plants by powderactivated carbon. *Water Sci. Technol.* **2012**, *66*, 2115. [CrossRef] [PubMed]

31. Clausen, K.; Lübken, M.; Pehl, B.; Bendt, T.; Wichern, M. Einsatz reaktivierter Aktivkohle von Wasserwerken zur Spurenstoffelimination in kommunalen Kläranlagen am Beispiel Düsseldorf (Use of reactivated activated carbon from water works for the elimination of trace substances in municipal sewage treatment plants using the example of Düsseldorf). *Korrespondenz Abwasser Abfall* **2014**, *61*, 1007–1012.

32. Grover, D.P.; Zhou, J.L.; Frickers, P.E.; Readman, J.W. Improved removal of estrogenic and pharmaceutical compounds in sewage effluent by full scale granular activated carbon: Impact on receiving river water. *J. Hazard. Mater.* **2011**, *185*, 1005–1011. [CrossRef] [PubMed]

33. Kovalova, L.; Siegrist, H.; von Gunten, U.; Eugster, J.; Hagenbuch, M.; Wittmer, A.; Moser, R.; McArdell, C.S. Elimination of Micropollutants during Post-Treatment of Hospital Wastewater with Powdered Activated Carbon, Ozone, and UV. *Environ. Sci. Technol.* **2013**, *47*, 7899–7908. [CrossRef] [PubMed]

34. Luo, Y.; Guo, W.; Ngo, H.H.; Nghiem, L.D.; Hai, F.I.; Zhang, J.; Liang, S.; Wang, X.C. A review on the occurrence of micropollutants in the aquatic environment and their fate and removal during wastewater treatment. *Sci. Total Environ.* **2014**, *473–474*, 619–641. [CrossRef] [PubMed]

35. Metzger, S.; Tjoeng, I.O.; Rößler, A.; Schwentner, G.; Rölle, R. Kosten der Pulveraktivkohleanwendung zur Spurenstoffelimination am Beispiel ausgeführter und in Bau befindlicher Anlagen (Cost of powdered activated carbon application for the elimination of trace substances using the example of selected plants under construction.). *Korrespondenz Abwasser Abfall* **2014**, *61*, 1029–1037.

36. Baresel, C.; Ek, M.; Harding, M.; Magnér, J.; Allard, A.S.; Karlsson, J. *Complementary Tests for a Resource Efficient Advanced Sewage Treatment*; Subreport SystemLäk Project; B2287; IVL Swedish Environmental Research Institute: Stockholm, Sweden, 2017.

37. Barillon, B.; Ruel, S.M.; Langlais, C.; Lazarova, V. Energy efficiency in membrane bioreactors. *Water Sci Technol.* **2013**, *67*, 2685–2691. [CrossRef] [PubMed]

38. Ioannou-Ttofa, L.; Foteinis, S.; Chatzisymeon, E.; Fatta-Kassinos, D. The environmental footprint of a membrane bioreactor treatment process through Life Cycle Analysis. *Sci. Total Environ.* **2016**, *568*, 306–318. [CrossRef] [PubMed]

39. Larrea, A.; Rambor, A.; Fabiyi, M. Ten years of industrial and municipal membrane bioreactor (MBR) systems–Lessons from the field. *Water Sci. Technol.* **2014**, *70*, 279–288. [CrossRef] [PubMed]

40. Meng, F.; Chae, S.-R.; Shin, H.-S.; Yang, F.; Zhou, Z. Recent Advances in Membrane Bioreactors: Configuration Development, Pollutant Elimination, and Sludge Reduction. *Environ. Eng. Sci.* **2012**, *29*, 139–160. [CrossRef]

41. Pinnekamp, J. Membrantechnik für die Abwasserreinigung (Membrane Technology for Wastewater Treatment). In *Siedlungswasser- und Siedlungsabfallwirtschaft Nordrhein-Westfalen*; Aktual, A., Ed.; FiW-Verl: Aachen, Germany, 2006.

42. Baresel, C.; Westling, K.; Samuelsson, O.; Andersson, S.; Royen, H.; Andersson, S.; Dahlén, N. Membrane Bioreactor Processes to Meet Todays and Future Municipal Sewage Treatment Requirements? *Int. J. Water Wastewater Treat.* **2017**, *3*. [CrossRef]

43. Lipp, P.; Kreißel, K.; Meuler, S.; Bischof, F.; Tiehm, A. Influencing parameters for the operation of an MBR with respect to the removal of persistent organic pollutants. *Desalin. Water Treat.* **2009**, *6*, 102–107. [CrossRef]

44. Radjenović, J.; Petrović, M.; Barceló, D. Fate and distribution of pharmaceuticals in wastewater and sewage sludge of the conventional activated sludge (CAS) and advanced membrane bioreactor (MBR) treatment. *Water Res.* **2009**, *43*, 831–841. [CrossRef] [PubMed]

45. Sipma, J.; Osuna, B.; Collado, N.; Monclús, H.; Ferrero, G.; Comas, J.; Rodriguez-Roda, I. Comparison of removal of pharmaceuticals in MBR and activated sludge systems. *Desalination* **2010**, *250*, 653–659. [CrossRef]

46. Ek, M.; Bergström, R.; Magnér, J.; Harding, M.; Baresel, C. *Aktivt kol för Avlägsnande av Läkemedelsrester ur Behandlat Avloppsvatten (Activated Carbon for Removal of Pharmaceutical Residues from Treated Wastewater)*; B2089; IVL Swedish Environmental Research Institute: Stockholm, Sweden, 2013.

47. Ek, M.; Bergström, R.; Baresel, C. *Avskiljning av Läkemedelsrester Med Granulerat Aktivt kol–Försök vid Himmerfjärdsverket (Removal of Pharmaceutical Residues with Granulated Activated Carbon)*; U4492; IVL Swedish Environmental Research Institute: Stockholm, Sweden, 2013.

48. Magnusson, K.; Jörundsdóttir, H.; Norén, F.; Lloyd, H.; Talvitie, J. *Microlitter in Sewage Treatment Systems: A Nordic Perspective on Waste Water Treatment Plants as Pathways for Microscopic Anthropogenic Particles to Marine Systems*; TemaNord, Nordic Council of Ministers: Copenhagen, Denmark, 2016.

49. Park, J.; Yamashita, N.; Park, C.; Shimono, T.; Takeuchi, D.M.; Tanaka, H. Removal characteristics of pharmaceuticals and personal care products: Comparison between membrane bioreactor and various biological treatment processes. *Chemosphere* **2017**, *179*, 347–358. [CrossRef]

50. El Gamal, M.; Mousa, H.A.; El-Naas, M.H.; Zacharia, R.; Judd, S. Bio-regeneration of activated carbon: A comprehensive review. *Sep. Purif. Technol.* **2018**, *197*, 345–359. [CrossRef]

51. Sbardella, L.; Comas, J.; Fenu, A.; Rodriguez-Roda, I.; Weemaes, M. Advanced biological activated carbon filter for removing pharmaceutically active compounds from treated wastewater. *Sci. Total Environ.* **2018**, *636*, 519–529. [CrossRef]

52. Baresel, C.; Cousins, A.P.; Hörsing, M.; Ek, M.; Ejhed, H.; Allard, A.S.; Magnér, J.; Westling, K.; Wahlberg, C.; Fortkamp, U.; et al. *Pharmaceutical Residues and Other Emerging Substances in the Effluent of Sewage Treatment Plants–Review on Concentrations, Quantification, Behaviour, and Removal Options*; B2226; IVL Swedish Environmental Research Institute: Stockholm, Sweden, 2015.

Article

Biodegradation of Picolinic Acid by *Rhodococcus* sp. PA18

Yanting Zhang [1], Junbin Ji [1], Siqiong Xu [1], Hongmei Wang [2], Biao Shen [3], Jian He [1], Jiguo Qiu [1,*] and Qing Chen [2,3,*]

[1] College of Life Sciences, Nanjing Agricultural University, Nanjing 210095, China; 2016116054@njau.edu.cn (Y.Z.); 2018216022@njau.edu.cn (J.J.); 2018116049@njau.edu.cn (S.X.); hejian@njau.edu.cn (J.H.)

[2] College of Life Sciences, Zaozhuang University, Zaozhuang 277100, China; zzxywhm@163.com (H.W.)

[3] College of Environmental and Resource Sciences, Nanjing Agricultural University, Nanjing 210095, China; shenbiao@njau.edu.cn

* Correspondence: qiujiguo@njau.edu.cn (J.Q.); cheniqing8686@126.com (Q.C.); Tel.: +86-632-3060356(Q.C.); Fax: +86-632-3786736 (Q.C.)

Received: 7 February 2019; Accepted: 6 March 2019; Published: 11 March 2019

Abstract: Picolinic acid (PA), a C2-carboxylated pyridine derivative, is a significant intermediate used in industrial production. PA is considered hazardous for the environment and human health. In this study, a Gram-positive bacterium, *Rhodococcus* sp. PA18, which aerobically utilizes PA as a source of carbon and energy, was isolated. The strain completely degraded 100 mg/L PA within 24 h after induction and formed 6-hydroxypicolinic acid (6HPA), a major PA metabolite, which was identified using ultraviolet-visible spectroscopy, high performance liquid chromatography, and liquid chromatography/time of flight-mass spectrometry analyses. The cell-free extracts converted the PA into 6HPA when phenazine methosulfate was used as an electron acceptor. To our knowledge, this is the first report showing that PA can be metabolized by *Rhodococcus*. In conclusion, *Rhodococcus* sp. PA18 may be potentially used for the bioremediation of environments polluted with PA.

Keywords: picolinic acid; biodegradation; *Rhodococcus*; 6-hydroxypicolinic acid

---

1. Introduction

Picolinic acid (PA), a pyridine derivative [1], has emerged as an important intermediate from the industrial syntheses of agricultural chemicals, drugs, dyestuffs, dyes, textiles, and mining [1–4]. It is also a dead-end product of L-tryptophan biosynthesis in living organisms [5–7]. In many common biological processes, PA is produced from the biodegradation of nitrobenzene, catechol, and anthranilic acid [8,9]. Owing to its hydrophilic nature (water solubility of 887 g/L at 20 °C), PA is easily transported to the aquatic environment and to soil [1,10]. Reports show that the pyridine concentration in wastewaters ranges from 20 to 300 mg/L [4]. Nevertheless, the toxicity or disadvantages of the use of PA have also been found. Owing to their relationship with environmental health, the removal of PA from contaminated ecosystems is considered essential for controlling environmental damage. Both physical and chemical methods are too expensive and ineffective in waste disposal management. Biological methods are efficient and cost effective, and hence, play an important role in the clean-up of toxic and hazardous wastes in the contaminated environment. Studies show that PA can be degraded efficiently by microorganisms, and many PA-utilizing bacterial strains, such as *Achromobacter* sp. JS18 [11], *Alcaligenes faecalis* JQ135 [12], *Arthrobacter picolinophilus* DSM 20665 [13], *Burkholderia* sp. ZD1 [3], and *Streptomyces* sp. Z2 [14], have been isolated and characterized.

An oxidative attack on the *N*-heterocyclic aromatic ring of PA is the main step initiating its bacterial degradation [1,3]. However, all of the above strains are Gram-negative bacteria, and whether Gram-positive bacteria can also degrade PA is not known. In this study, a Gram-positive PA-assimilating strain, *Rhodococcus* sp. PA18, was isolated, identified, and characterized. The aerobic degradation of PA by strain PA18 and the formation of 6-hydroxypicolinic acid (6HPA), the initial product formed after the oxidative attack on the *N*-heterocyclic aromatic ring, have been discussed.

To the best of our knowledge, this is the first report on the degradation of PA by a Gram-positive bacterium, *Rhodococcus* sp. PA18. This may be developed as a potentially low cost and environmentally-friendly approach to restore the environments contaminated by PA.

2. Materials and Methods

2.1. Chemicals and Media

PA and 6HPA were acquired from J&K Scientific Ltd. (Shanghai, China). Phenazine methosulfate (PMS), methylene blue (MB), and 2,6-dichloroindophenol (DCIP) were purchased from Sangon Biotech Co., Ltd. (Shanghai, China). All of the other chemicals and solvents used in this study were available commercially.

The Luria Bertani (LB) medium contained (per liter) tryptone (10.0 g), yeast extract (5.0 g), and NaCl (10.0 g). The mineral salts medium (MSM) contained (per liter) $(NH_4)_2SO_4$ (1.0 g), NaCl (1.0 g), K_2HPO_4 (1.5 g), KH_2PO_4 (0.5 g), $MgSO_4$ (0.2 g), and 1 mL trace elements medium stock solution (0.13 g $MnSO_4 \cdot H_2O$, 0.23 g $ZnCl_2$, 0.03 g $CuSO_4 \cdot H_2O$, 0.42 g $CoCl_2 \cdot 6H_2O$, 0.15 g $Na_2MoO_4 \cdot 2H_2O$, and 0.05 g $AlCl_3 \cdot 6H_2O$). The media were adjusted to pH 7.0 using HCl or NaOH, and were autoclaved at 121 °C for 30 min. For the solid media, 1.5% agar (w/v) was added.

2.2. Isolation and Identification of PA-Degrading Bacteria

Soil samples were collected from a farmland in Nanjing City, China. The enrichment culture was initiated with 5.0 g of soil samples in 100 mL MSM, containing 500 mg/L PA as the sole carbon source. It was incubated under aerobic conditions in a rotary shaker at 30 °C and 150 rpm for one week, and then 5% (v/v) of the culture was transferred to fresh MSM containing PA. The process was repeated four times. The culture was diluted and spread on MSM agar plates containing PA (500 mg/L). The colonies were tested for their PA degradation capability. A strain named PA18, which showed a high PA degrading efficiency, was selected for further studies. The morphological and physiological characteristics of PA18 were characterized, and its 16S rDNA sequence was analyzed according to the method described by Nie et al. [15].

2.3. Biodegradation of PA by a Rhodococcus Strain

The strain *Rhodococcus* sp. PA18 was cultured in 100 mL of liquid LB medium for 24 h, until the late log phase. The cells were harvested via centrifugation at 5000 rpm for 10 min, washed twice with fresh MSM, and then transferred into MSM containing 100 mg/L PA. After seven days, the *Rhodococcus* cells were harvested, washed twice with fresh MSM, and then resuspended in fresh MSM. The optical density of the cells at 600 nm (OD_{600}) was adjusted to 2.0. These PA-pre-cultured cells were used for further biodegradation assays. The ability of PA18 to degrade and utilize PA was assessed in 250 mL-conical flasks, with 100 mL of MSM supplemented with a constant initial concentration of PA (100 mg/L), and the cells were incubated in a rotary shaker at 30 °C and 150 rpm. The experiment was repeated in triplicate and the control experiment was set up as above, but without the inoculation of the bacteria. The samples were collected periodically from the cultures. The growth was monitored by measuring the OD_{600}, and the residual PA concentration was measured using HPLC, as described below.

2.4. Effects of Different Factors on PA Biodegradation

In order to determine the optimal conditions for PA biodegradation, different inoculum sizes and initial PA concentrations were tested. Inoculum sizes of 0.005, 0.200, 0.500, and 1.000 at OD_{600}

were used when the initial PA concentration was 100 mg/L. Conversely, the initial concentration of PA varied from 50, 100, 200, 300, 400, to 500 mg/L when the inoculum size was maintained at 0.005 at OD_{600}. All of the above experiments were conducted in 250 mL flasks containing 100 mL MSM in a rotary shaker at 30 °C and 150 rpm. Each experiment was conducted in triplicate, and the residual PA concentration was measured by HPLC, as described below.

2.5. Identification of Metabolites during PA Degradation

To identify the metabolites during PA degradation, the strain PA18 was inoculated in a MSM medium with PA (100 mg/L), and was incubated at 30 °C and 150 rpm. The samples were collected at appropriate intervals for further analysis using UV-VIS spectroscopy, high performance liquid chromatography (HPLC), and liquid chromatography/time of flight-mass spectrometry (LC/TOF-MS).

2.6. Cell-Free Extract Activity Assays

The PA-pre-cultured PA18 cells were grown to the mid-log phase in MSM with 100 mg/L PA. After centrifugation (4 °C, 12,000 rpm, 10 min), the harvested cells were resuspended in a 50 mM phosphate buffer (pH 7.0). The cell-free extract was prepared by sonication in an ice-water bath for 30 min (1 s with 2 s intervals). After sonication, the supernatant liquid was separated from the cell debris by centrifugation at 12,000 rpm for 30 min. The supernatant was collected for enzyme assays. The protein concentrations were measured using the Bradford method [16,17]. The picolinic acid hydroxylase activity of the cell-free extract was determined by measuring the increase in the absorbance of the reaction mixture at 310 nm, the absorption maximum of the product 6HPA (ε = 4.45 cm^{-1} mM^{-1}). The reaction mixture contained 0.1 mM PA and a 0.1 mM electron acceptor in 500 μL of cell extract. The reaction was started by the addition of PA. To evaluate the molar ratio among PA, 6HPA, and PMS, the PMS was maintained at 0.1 mM, while the concentration of PA was varied from 0.00 mM to 0.50 mM. One unit of activity was defined as the amount of enzyme that catalyzed the formation of 1 μmol 6HPA in 1 min. Specific activity refers to the number of units of the enzyme present per milligram protein. The kinetic data were evaluated using nonlinear regression analysis with the Michaelis–Menten equation. All of the data were collected from three independent determinations.

2.7. Analytical Methods

The cell density was estimated spectrophotometrically by detecting the absorbance at 600 nm using an UV-VIS spectrophotometer (Shimadzu, UV-2450, Kyoto, Japan). The concentration of PA and its catabolic intermediates in the supernatants was primarily monitored using an UV-VIS spectrophotometer (Shimadzu, UV-2450, Kyoto, Japan). The spectral data were collected from 400 nm to 200 nm. PA and 6HPA were identified using an HPLC analysis on a Shimadzu AD20 system equipped with a C_{18} reverse phase column (250 × 4.60 mm, 5 μm; Agilent Technologies, Santa Clara, CA, USA). The concentrations of the compounds were calculated using standard samples. The detection wavelength was set at 260 and 310 nm. The mobile phase was a mixture of methanol–water (10:90, vol/vol) with 0.2% formic acid, and the flow rate was 0.7 mL/min. The column temperature was 30 °C. The LC/TOF-MS analysis was performed in a TripleTOF 5600 (AB SCIEX) mass spectrometer [18], and the conditions were identical to that used for HPLC.

3. Results

3.1. Isolation and Identification of the PA-Degrading Strains

Several strains demonstrated a PA utilization from the enriched soil solution after four weeks. Among these strains, one bacterium, designated as PA18, showed a high PA-degrading efficiency. The strain grew in the presence of 6.5% (*w/v*) NaCl, and the D-Xylose, cellobiose, lactose, and D-mannitol could be used as the sole source of the carbon and energy. A phylogenetic tree was constructed based on a BLAST analysis of its 16S rDNA sequence (GenBank accession number: MH

681664). The strain PA18 showed a 99%–100.0% sequence similarity to *R. erythropolis* NBRC 15567^T (BCRM01000055), *R. degradans* CCM 4446^T (JQ776649), and *R. qingshengii* JCM 15477^T (LRRJ01000016), and formed a subclade with *R. erythropolis* NBRC 15567^T (Figure 1). Thus, it was identified as a member of *Rhodococcus* sp. The genus *Rhodococcus* is considered one of the most promising groups of organisms suitable for the biodegradation of compounds that cannot be easily transformed by other organisms [19]. Members of the genus *Rhodococcus* exist as abundant indigenous bacterial communities in localities contaminated with various aromatic pollutants [20,21], indicating that *Rhodococcus* spp. may play an important role in bioremediation. Numerous *Rhodococcus* have been isolated, for example, *R. pyridinivorans* PDB9 and *Rhodococcus* spp. have been reported to degrade pyridine [22,23], *Rhodococcus* sp. Y22 and *R. rhodochrous* J1 degrade nicotine or nicotinic acid [24,25], and *Rhodococcus* sp. B1 degrades quinoline or related-compounds [26]. However, no *Rhodococcus* species has been shown to degrade PA to date. This study provided new insights regarding PA catabolism and biodegradation by the *Rhodococcus* species, which may be used for the bioremediation of polluted environments.

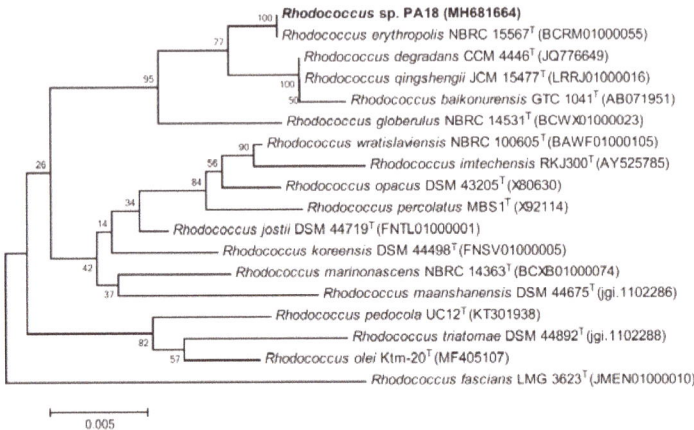

Figure 1. Neighbor-joining phylogenetic tree shows the position of strain picolinic acid 18 (PA18) among some members of the genus *Rhodococcus*, based on the 16S rDNA sequences.

3.2. Growth of Strain PA18 and PA Biodegradation

The relationship between the growth of strain PA18 and PA degradation in MSM is indicated by Figure 2. The concentration of PA remained unchanged in the control samples, which did not contain strain PA18. This confirmed that abiotic losses, such as the volatilization and adsorption of PA, were negligible in this study. PA18 degraded 100 mg/L PA at a relatively higher rate within the first 24 h (Figure 2A). The degradation rate decreased substantially from 24 to 28 h, possibly as a result of a lack of substrate, with complete degradation within 28 h. Furthermore, the PA degradation was associated with a concomitant increase in the bacterial cell density. The OD$_{600}$ value reached 0.31 within 16 h. Interestingly, after a four-day lag phase, the strain PA18 cultured in a LB medium also showed a high-efficiency degradation of PA (Figure 2B). In contrast, the negative control with no inoculated PA18 showed no discernable degradation of PA over the same 120 h. At the same time, the cells grew rapidly with the degradation of PA.

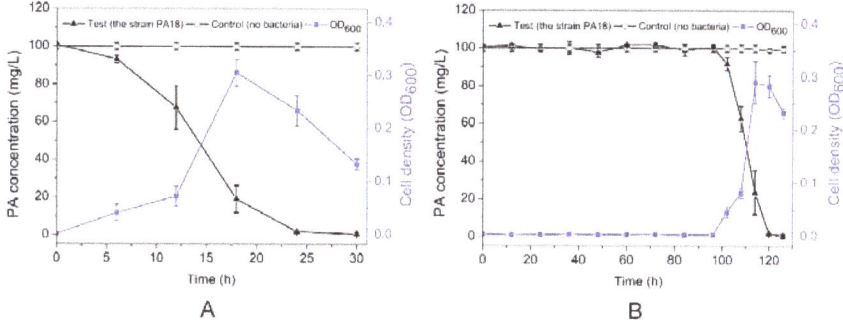

Figure 2. Utilization of PA during the growth of *Rhodococcus* sp. PA18 in mineral salts medium (MSM) (pH 7.0). The strain was pre-cultured with PA (**A**). The strain was pre-cultured in a Luria Bertani (LB) medium (**B**).

3.3. Effects of Initial PA Concentration and Inoculum of the Strain on PA Biodegradation

To investigate the effect of the environmental factors on PA biodegradation, the effects of the initial PA concentration and the inoculum of the strain on the PA degradation rate were investigated. As shown in Figure 3A, the PA degradation efficiencies were 10.4%, 28.6%, 68.0%, and 95.9% after 6-h of incubation, from a starting OD_{600} of 0.05%, 0.20%, 0.50%, and 1.00%, respectively, when the initial PA concentration was 100 mg/L at 30°C, pH 7.0. Furthermore, as shown in Figure 3B, when the initial concentration of PA was increased from 50 mg/L to 300 mg/L, the time required for complete PA removal increased from 18 to 48 h, which increased to 84 h when the initial concentration of PA was increased to 400 mg/L. However, the PA could not be degraded when the concentration was 500 mg/L. One possible reason was that the excessive concentration of PA was toxic for the strain. The delay period was extended as the initial PA concentration increased. Compared with other PA-degrading strains previously reported, such as, *Achromobacter sp.* JS18 [11], *Alcaligenes faecalis* JQ135 [12], *Arthrobacter picolinophilus* DSM 20665 [13], *Burkholderia* sp. ZD1 [3], and *Streptomyces* sp. Z2 [14], strain PA18 possessed a very high degradation efficiency.

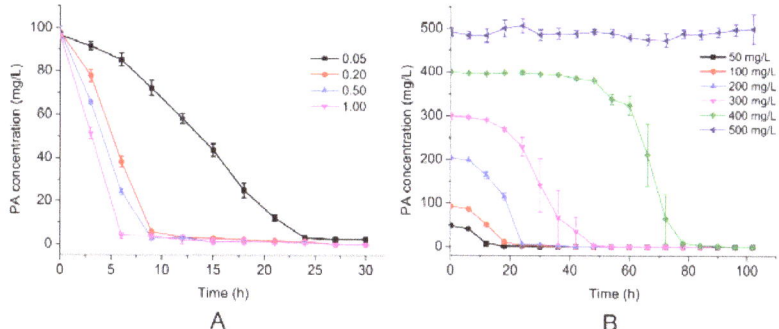

Figure 3. Effects of inocula (**A**) and the initial PA concentration (**B**) on the degradation of PA in MSM. The error bars show standard deviations.

3.4. Identification of PA Metabolites

PA degradation by PA18 was analyzed preliminarily using ultraviolet-visible spectroscopy (UV-VIS). In addition to the parent compound, PA (maximum at 260 nm), one major metabolite was detected with an increase in the absorption at 310 nm (Figure 4A). The intermediate product was collected and subjected to high performance liquid chromatography (HPLC). The peak corresponding to PA was detected at a retention time of 4.1 min, and an unknown peak appeared at 8.3 min (Figure 4B).

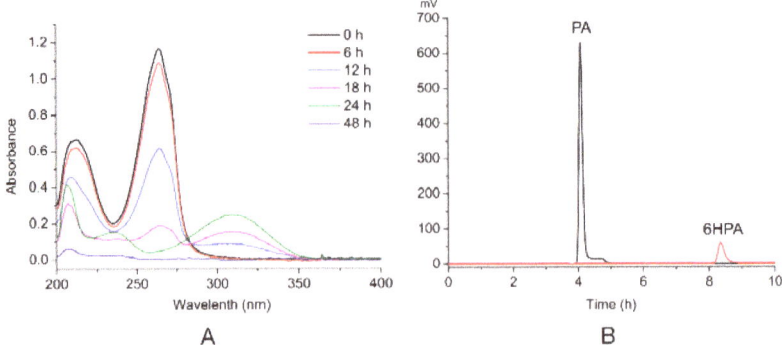

Figure 4. PA degradation analysis by the strain PA18 using UV-VIS and high-performance liquid chromatography (HPLC). (**A**) The UV-VIS absorption of each compound during PA degradation from 200 to 400 nm. (**B**) HPLC analysis of the conversion of PA into 6HPA at 12 h. The black line indicates the sample collected at 0 h and the red line indicates at 12 h.

Then, the sample was further analyzed using liquid chromatography/time of flight-mass spectrometry (LC/TOF-MS). The fragment at m/z 124.0398 [M+H]$^+$ conformed to the molecular weight of 124.0354 for PA (Figure 5). In addition, another molecular ion of m/z 140.0346 [M+H]$^+$ was detected. An increment of 16 in the molecular weight indicated that an oxygen atom was added to the PA, which was consistent with the molecular weight of the predicted metabolite 6HPA ($C_6H_5NO_3$, m/z 140.0303) (Figure 5). Therefore, the PA was possibly hydroxylated by PA18, which was consistent with the results of the previous reports on PA degradation by *Burkholderia* sp. ZD1 [3] and *Alcaligenes faecalis* JQ135 [12]. A standard sample of 6HPA was also analyzed using HPLC under the same conditions, and its retention time was identical to that of the unknown peak mentioned above. Therefore, the intermediate metabolite was identified as 6HPA. Throughout the degradation process, no other intermediates were detected.

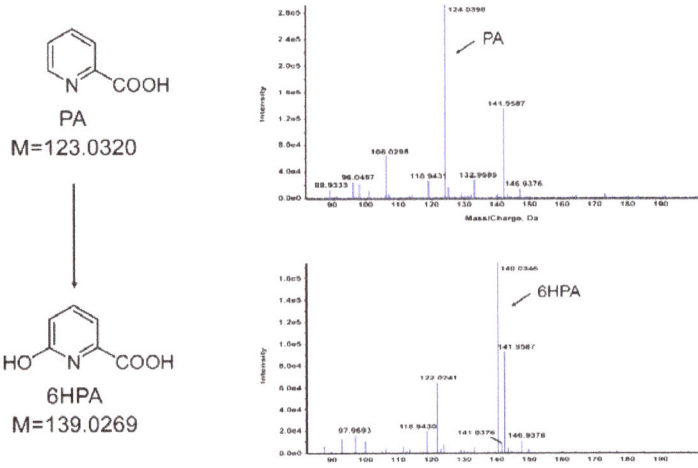

Figure 5. LC/TOF-MS profiles of PA and 6HPA.

Usually, the first step in the microbial degradation of the aromatic ring or heterocyclic aromatic compounds is the introduction of a hydroxyl group, and these dehydroxylated substrates can be subsequently decarboxylated [27]. Finally, it can be easily cleaved by dioxygenase or mooxygenase via either an ortho-cleavage pathway or a meta-cleavage pathway, leading to the formation of tricarboxylic acid cycle intermediates [28,29]. The first step, hydroxylation of pyridine or its derivatives, is the key step in biodegradation. Several studies have already shown that pyridine [30], PA [12,31], nicotine [32], and quinolone [33] are hydroxylated at the carbon adjacent to the heteroatom. Pyridine derivatives possessing hydroxyl groups can be used as important starting materials for the synthesis of agrochemicals and pharmaceuticals [34,35]. Future research will be focused on cloning the gene encoding hydroxylase, as well as further study of the enzymatic characteristics.

3.5. Activity of the Enzyme in the Cell Extract

As shown in Figure 6A, a new absorption peak at 310 nm appeared, the intensity of which increased significantly, when PMS acted as an electron acceptor. This corresponded to the absorption maximum of the product 6HPA. However, no 6HPA was produced without the electron acceptor or when the electron acceptor was MB or DCIP (data not shown). These results indicate that the cell extract was able to convert PA into 6HPA, with PMS as the electron acceptor. The Km of the crude enzyme for PA at pH 7.0 and 30°C was 68.86 ± 14.85 μM, and the specific activity was 154.17 U/mg. To verify the quantitative relationship between the PA and PMS, the amount of PA was changed continuously from 0.0 to 0.5 mM, whereas that of PMS was maintained at a constant of 0.1 mM, as shown in Figure 6B. The amount of product 6HPA increased with the concentration of PA, and the PA:6HPA ratio was 1:1. Subsequently, the amount of 6HPA formed did not change, even when the substrate (PA) concentration was increased, indicating that the optimal molar ratio of PA:PMS was 1:1. The hydroxylases of aromatic or N-heterocyclic aromatic compounds that incorporate oxygen into the product are Rieske non-heme iron aromatic ring-hydroxylating oxygenase or cytochrome P450s [29,36]. In strain *Pseudomonas putida* KT2440, the nicotinic acid can be converted to 6-hydroxynicotinic acid by a two-component hydroxylase (NicAB), whose electron transport chain to the molecular oxygen includes a cytochrome c domain [37]. Interestingly, unlike other studies, the hydroxylase in the strain PA18 cell extract that can convert PA to 6HPA uses PMS as an electron acceptor.

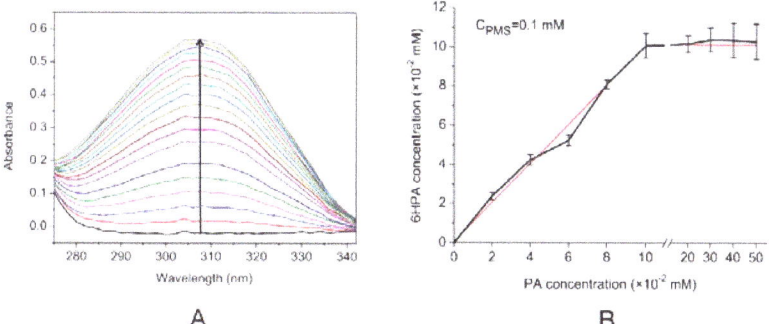

Figure 6. Measurement of the enzyme activities in cell-free extracts. (**A**) Spectrophotometric changes during the transformation of PA. The spectra were recorded every 1 min. The arrow indicates the directions of the spectral changes. (**B**) Relationship between PA and 6HPA while the concentration of PMS was constant. PA concentration was increased from 0.00 to 0.50 mM whereas the phenazine methosulfate (PMS), concentration was constant at 0.10 mM. The data were calculated from three independent replicates, and error bars indicate standard deviations.

4. Conclusions

In this study, a novel bacterial species, *Rhodococcus* sp. PA18, was isolated and demonstrated to possess an outstanding PA biodegradation efficiency, even at a low initial inoculum size. This strain completely degraded 100 mg/L PA within 24 h after prior induction. UV-VIS spectroscopy, HPLC, and LC/TOF-MS were used to deduce the biodegradation pathway, and 6HPA was identified as one intermediate. Furthermore, in the crude enzymology experiment, the optimal molar ratio between PA and PMS was 1:1. In addition, the quantitative relationship between the substrate (PA) and the product (6HPA) was 1:1. This is the first study to identify the PA-degrading ability of *Rhodococcus*. Further studies are required in order to reveal the biochemical process and mechanisms involved in the associated biodegradation pathway.

Author Contributions: conceived and designed the experiments, J.Q., Q.C., and J.H.; performed the experiments, Y.Z., J.J., S.X., and H.M; resources, J.Q., Q.C., and J.H.; writing (original draft preparation), Y.Z. and Q.C.; writing (review and editing), J.Q., B.S., and J.H.; funding acquisition, Q.C. and J.Q.

Funding: This research was funded by the Natural Science Foundation of China (no. 31600080 and no. 31870092) and the Natural Science Foundation of Shandong Province, China (ZR2016CB29).

Conflicts of Interest: The authors declare no conflict of interest.

References

1. Kaiser, J.P.; Feng, Y.; Bollag, J.M. Microbial metabolism of pyridine, quinoline, acridine, and their derivatives under aerobic and anaerobic conditions. *Microbiol. Rev.* **1996**, *60*, 483–498. [PubMed]
2. Yao, Y.; Tang, H.; Ren, H.; Yu, H.; Wang, L.; Zhang, W.; Behrman, E.J.; Xu, P. Iron(II)-dependent dioxygenase and *N*-formylamide deformylase catalyze the reactions from 5-hydroxy-2-pyridone to maleamate. *Sci. Rep.* **2013**, *3*, 3235. [CrossRef] [PubMed]
3. Zheng, C.; Wang, Q.; Ning, Y.; Fan, Y.; Feng, S.; He, C.; Zhang, T.C.; Shen, Z. Isolation of a 2-picolinic acid-assimilating bacterium and its proposed degradation pathway. *Bioresour. Technol.* **2017**, *245*, 681–688. [CrossRef] [PubMed]
4. Huang, D.; Liu, W.; Wu, Z.L.; Liu, G.M.; Yin, H.; Chen, Y.R.; Hu, N.; Jia, L. Removal of pyridine from its wastewater by using a novel foam fractionation column. *Chem. Eng. J.* **2017**, *321*, 151–158. [CrossRef]
5. Heyes, M.P.; Eugene, O.; Saito, K. Different kynurenine pathway enzymes limit quinolinic acid formation by various human cell types. *Biochem. J.* **1987**, *326*, 351–356. [CrossRef]
6. Bryleva, E.Y.; Brundin, L. Kynurenine pathway metabolites and suicidality. *Neuropharmacology* **2017**, *112*, 324–330. [CrossRef] [PubMed]
7. Esquive, D.G.; Ramirez-Ortega, D.; Pineda, B.; Castro, N.; Rios, C.; de la Cruz, V.P. Kynurenine pathway metabolites and enzymes involved in redox reactions. *Neuropharmacology* **2017**, *112*, 331–345.
8. Asano, Y.; Yamamoto, Y.; Yamada, H. Catechol 2, 3-dioxygenase-catalyzed synthesis of picolinic acids from catechols. *Biosci. Biotechnol. Biochem.* **1994**, *58*, 2054–2056. [CrossRef]
9. Nishino, S.F.; Spain, J.C. Degradation of nitrobenzene by a *Pseudomonas pseudoalcaligenes*. *Appl. Environ. Microbiol.* **1993**, *59*, 2520–2525. [PubMed]
10. Tian, Y.Q.; Li, W.H. A new process of synthesis of chromium-2-picolinate by chromic anhydride oxidation. *Chem. Eng.* **2005**, *19*, 53–54. (In Chinese)
11. Kutanovas, S.; Karvelis, L.; Vaitekūnas, J.; Stankevičiūtė, J.; Gasparavičiūtė, R.; Meškys, R. Isolation and characterization of novel pyridine dicarboxylic acid-degrading microorganisms. *Chemija* **2016**, *30*, 74–83.
12. Qiu, J.G.; Zhang, J.J.; Zhang, Y.T.; Wang, Y.H.; Tong, L.; Hong, Q.; He, J. Biodegradation of picolinic acid by a newly isolated bacterium *Alcaligenes faecalis* strain JQ135. *Curr. Microbiol.* **2017**, *74*, 508–514. [CrossRef] [PubMed]
13. Siegmund, I.; Koenig, K.; Andreesen, J.R. Molybdenum involvement in aerobic degradation of picolinic acid by *Arthrobacter picolinophilus*. *FEMS Microbiol. Lett.* **1990**, *67*, 281–284. [CrossRef]
14. Zheng, C.; Zhou, J.; Wang, J.; Qu, B.; Lu, H.; Zhao, H. Aerobic degradation of 2-picolinic acid by a nitrobenzene-assimilating strain: *Streptomyces* sp. Z2. *Bioresour. Technol.* **2009**, *100*, 2082–2084. [CrossRef] [PubMed]

15. Nie, Z.J.; Hang, B.J.; Cai, S.; Xie, X.T.; He, J.; Li, S.P. Degradation of cyhalofop-butyl (CyB) by *Pseudomonas azotoformans* strain QDZ-1 and cloning of a novel gene encoding CyB-hydrolyzing esterase. *J. Agric. Food Chem.* **2011**, *59*, 6040–6046. [CrossRef] [PubMed]

16. Bradford, M.M. A rapid and sensitive method for the quantitation of microgram quantities of protein utilizing the principle of protein-dye binding. *Anal. Biochem.* **1976**, *72*, 248–254. [CrossRef]

17. Qiu, J.G.; Liu, B.; Zhao, L.; Zhang, Y.T.; Cheng, D.; Yan, X.; Jiang, J.D.; Hong, Q.; He, J. A novel degradation mechanism for pyridine derivatives in *Alcaligenes faecalis* JQ135. *Appl. Environ. Microbiol.* **2018**, *84*, e00910-18. [CrossRef] [PubMed]

18. Yang, Z.G.; Jiang, W.K.; Wang, X.H.; Cheng, T.; Zhang, D.S.; Wang, H.; Qiu, J.G.; Cao, L.; Hong, Q. An amidase gene ipaH is responsible for the initial degradation step of iprodione in strain *Paenarthrobacter* sp. YJN-5. *Appl. Environ. Microbiol.* **2018**, *84*, e01150-18. [CrossRef] [PubMed]

19. Warhurst, A.M.; Fewson, C.A. Biotransformations catalyzed by the genus *Rhodococcus*. *Crit. Rev. Biotechnol.* **1994**, *14*, 29–73. [CrossRef]

20. Fahy, A.; McGenity, T.J.; Timmis, K.N.; Ball, A.S. Heterogeneous aerobic benzene-degrading communities in oxygen-depleted groundwaters. *FEMS Microbiol. Ecol.* **2006**, *58*, 260–270. [CrossRef] [PubMed]

21. Leigh, M.B.; Prouzová, P.; Macková, M.; Macek, T.; Nagle, D.P.; Fletcher, J.S. Polychlorinated biphenyl (PCB)-degrading bacteria associated with trees in a PCB-contaminated site. *Appl. Environ. Microbiol.* **2006**, *72*, 2331–2342. [CrossRef] [PubMed]

22. Lee, S.T.; Lee, S.B.; Park, Y.H. Characterization of a pyridine-degrading branched Gram-positive bacterium isolated from the anoxic zone of an oil shale column. *Appl. Microbiol. Biotechnol.* **1991**, *35*, 824–829. [CrossRef]

23. Yoon, J.H.; Kang, S.S.; Cho, Y.G.; Lee, S.T.; Kho, Y.H.; Kim, C.J.; Park, Y.H. *Rhodococcus pyridinivorans* sp. nov., a pyridine-degrading bacterium. *Int. J. Syst. Evol. Microbiol.* **2000**, *50*, 2173–2180. [CrossRef] [PubMed]

24. Mathew, C.D.; Nagasawa, T.; Kobayashi, M.; Yamada, H. Nitrilase-catalyzed production of nicotinic acid from 3-cyanopyridine in *Rhodococcus rhodochrous* J1. *Appl. Environ. Microbiol.* **1988**, *54*, 1030–1032. [PubMed]

25. Gong, X.W.; Ma, G.H.; Duan, Y.Q.; Zhu, D.L.; Chen, Y.K.; Zhang, K.Q.; Yang, J.K. Biodegradation and metabolic pathway of nicotine in *Rhodococcus* sp. Y22. *World J. Microbiol. Biotechnol.* **2016**, *32*, 188. [CrossRef] [PubMed]

26. Schwarz, G.; Bauder, R.; Speer, M.; Rommet, T.O.; Lingens, F. Microbial metabolism of quinoline and related compounds. II. Degradation of quinoline by *Pseudomonas fluorescens* 3, *Pseudomonas putida* 86 and *Rhodococcus spec.* B1. *Biol. Chem. Hoppe Seyler* **1989**, *370*, 1183–1190. [CrossRef] [PubMed]

27. Yoshida, T.; Nagasawa, T. Enzymatic functionalization of aromatic N-heterocycles: Hydroxylation and carboxylation. *J. Biosci. Bioeng.* **2000**, *89*, 111–118. [CrossRef]

28. Mallick, S.; Chakraborty, J.; Dutta, T.K. Role of oxygenases in guiding diverse metabolic pathways in the bacterial degradation of low-molecular-weight polycyclic aromatic hydrocarbons: A review. *Crit. Rev. Microbiol.* **2011**, *37*, 64–90. [CrossRef] [PubMed]

29. Ghosal, D.; Ghosh, S.; Dutta, T.K.; Ahn, Y. Current state of knowledge in microbial degradation of polycyclic aromatic hydrocarbons (PAHs): A review. *Front. Microbiol.* **2016**, *7*, 1369. [CrossRef] [PubMed]

30. Watson, G.K.; Cain, R.B. Microbial metabolism of the pyridine ring. Metabolic pathways of pyridine biodegradation by soil bacteria. *Biochem. J.* **1975**, *146*, 157–172. [CrossRef] [PubMed]

31. Orpin, C.G.; Knight, M.; Evans, W.C. The bacterial oxidation of picolinamide, a photolytic product of diquat. *Biochem. J.* **1972**, *127*, 819–831. [CrossRef] [PubMed]

32. Qiu, J.G.; Ma, Y.; Wen, Y.Z.; Chen, L.S.; Wu, L.F.; Liu, W.P. Functional identification of two novel genes from *Pseudomonas* sp. strain HZN6 involved in the catabolism of nicotine. *Appl. Environ. Microbiol.* **2012**, *78*, 2154–2160. [CrossRef] [PubMed]

33. Shukla, O.P. 8-Hydroxycoumarin: An intermediate in the microbial transformation of quinoline. *Curr. Sci.* **1984**, *53*, 1145–1147.

34. Yasuda, M.; Sakamoto, T.; Sashida, R.; Ueda, M.; Morimoto, Y.; Nagasawa, T. Microbial hydroxylation of 3-cyanopyridine to 3-cyano-6-hydroxypyridine. *Biosci. Biotechnol. Biochem.* **1995**, *59*, 572–575. [CrossRef]

35. Yu, H.; Tang, H.Z.; Xu, P. Green strategy from waste to value-added-chemical production: Efficient biosynthesis of 6-hydroxy-3-succinoyl-pyridine by an engineered biocatalyst. *Sci. Rep.* **2014**, *4*, 5397. [CrossRef] [PubMed]

36. Hannemann, F.; Bichet, A.; Ewen, K.M.; Bernhardt, R. Cytochrome P450 systems-biological variations of electron transport chains. *BBA-Gen Subj.* **2007**, *1770*, 330–344. [CrossRef] [PubMed]

37. Jiménez, J.I.; Canales, Á.; Jiménez-Barbero, J.; Ginalski, K.; Rychlewski, L.; García, J.L.; Díaz, E. Deciphering the genetic determinants for aerobic nicotinic acid degradation: The *nic* cluster from *Pseudomonas putida* KT2440. *Proc. Natl. Acad. Sci. USA* **2008**, *105*, 11329–11334. [CrossRef] [PubMed]

Article

Chitosan Microbeads as Supporter for *Pseudomonas putida* with Surface Displayed Laccases for Decolorization of Synthetic Dyes

Zhiqiang Bai [1,2], Xiaowen Sun [1], Xun Yu [1] and Lin Li [1,*]

[1] State Key Laboratory of Agricultural Microbiology, College of Life Science and Technology,
    Huazhong Agriculture University, Wuhan 430070, China; baizq1987@126.com (Z.B.);
    sunxiaowen525@webmail.hzau.edu.cn (X.S.); yuxun@webmail.hzau.edu.cn (X.Y.)
[2] School of Life Sciences, Shanxi Datong University, Datong 037009, China
*   Correspondence: lilin@mail.hzau.edu.cn; Tel.: +86-27-8728-6952

Received: 3 December 2018; Accepted: 24 December 2018; Published: 3 January 2019

Abstract: Various untreated wastewaters contaminated with industrial dyes pose significant pollution hazards to the natural environment as well as serious risks to public health. The current study reports a new material with a configurative chitosan matrix and engineered *Pseudomonas putida* cells with surface-displayed laccases that can decolorize five industrial dyes. Through a self-configuring device, five chitosan microbeads (CTS-MBs) with different particle sizes were prepared. *P. putida* cells were then immobilized onto the CTS-MBs under optimized immobilization conditions, forming a degrading-biosorbent dual-function decolorization complex. Scanning electron microscope and infrared analysis confirmed the successful immobilization of the cells onto the CTS-MB matrix. The optimized CTS-MB1 with surface-grafted aldehyde groups (aCTS-MB1) complex was capable of decolorizing Acid Green 25 and Acid Red 18 over a pH range of 2.5–8.5 and a relatively broad temperature range of 15–85 °C, with a maximum relative decolorization value of over 94%; the complex was also able to efficiently decolorize Direct Red 243, Reactive Blue 220 and Reactive Blue 198. Moreover, the aCTS-MB1 composite showed favorable activity in continuous and regenerative decolorization reactions. Therefore, the chitosan-immobilized decolorizing material, with both improved mechanical strength and performance, shows potential for further large-scale or continuous processes.

Keywords: chitosan; *Pseudomonas putida*; immobilization; dye decolorization; degradation; biosorption

---

1. Introduction

Textile dyes comprise a large class of chemicals with highly comprehensive components, diverse origins and extensive applications and they constitute the main source of pollution in sewage discharge from the textile and printing industry. In China and Southeast Asia, contamination caused by various textile dyes from untreated industrial effluents is a serious environmental threat, as a variety of textile dyes are toxic or cross-coupled to toxic components that are relatively recalcitrant to degradation due to their complicated molecular structure [1,2]. Even a small amount of trace dye in water can seriously affect the transparency and gas solubility of the water [3]. Remediation of dye contamination often involves chemical, physical and various biological processes [4,5]. Over the past decade, intensive efforts have been made to develop effective bioprocesses for the treatment of dyes in wastewater effluents or other environments [2,6]. Of the various available biological decolorizers, the macromolecular polymer chitosan (CTS) has particular appeal due to its relatively high natural dye biosorption efficiency and its ease of transformation into biocomposites through physicochemical modification, which leads to efficient decolorization of various dyes [7–9].

CTS is a natural poly-amino glucosamine polymer that is widely distributed in the exoskeleton of various crustaceans, in the exo- and endocuticle of insects and in the cell wall of fungi, where it functions as an integral component as well as a mechanically strong supporting scaffold material for these organisms; moreover, CTS has been recognized as one of the most plentiful renewable organic resources worldwide [10,11]. The distinctive biological properties of CTS, particularly its biocompatibility, biodegradability, environmental friendliness and regenerability, as well as its relative permeability and cost-effectiveness, indicate its potential for applications in a variety of industrial fields, such as biochemical engineering, wastewater remediation, food processing [12,13], bone tissue engineering and other therapeutic applications [14,15].

In addition to biosorption, enzymatic biodegradation provides another promising approach to the decolorization of wastewater dyes [16]. Among various degrading enzymes such as phenoloxidases, laccases and tyrosinases, microbial laccases are being increasingly investigated as an effective and environmentally friendly means of treating industrial phenolic substrates [17]. Laccases (benzenediol:dioxygen oxidoreductases, EC1.10S.3.2) are a large group of multi-copper enzymes that are involved in different biological processes of organisms, such as lignifications in plants; morphogenesis, pathogenesis and detoxification in fungi; cuticle osteosis; and resistance to heave metals, chlorides, ultraviolet (UV) radiation and H_2O_2 in bacteria, among others [16,18,19]. Many previous investigations have demonstrated the ability of microbial laccases to oxidatively degrade a broad range of organic compounds, particularly aromatic substrates [20–22], including industrial or wastewater dyes [17,23,24]. In a similar approach, we developed a *Pseudomonas* cell surface display system to efficiently decolorize the anthraquinone dye Acid Green 25 and diazo dye Acid Red 18 [25] in which a mutated bacterial laccase (WlacD) [26] was projected onto the surface of target cells, enabling fast and goal-oriented decolorization on the cell surface. This system has proven advantages over freely suspended laccase and bacterial cells alone, including the elimination of mass transfer limitation, minimization of dye toxicity to living cells and particularly, promotion of the reaction rate. However, one technical drawback remained in this system: its inability to be applied in large-scale or continuous processes, as high loading of this type of biomass, which is composed of cells, will become clogged under continuous-flow conditions in a reactor. In addition, as a recombinant bacterial strain, carrying an antibiotic marker is also an important concern for environmental release. Therefore, further studies to improve the performance of this system for large-scale or continuous operations are required.

Pseudomonas putida is a well-known non-pathogenic and robust bacterium, with a wealth of oxidoreductases and a versatile metabolism capable of utilizing a wide range of inorganic and organic compounds [27,28]. Moreover, this bacterium exhibits a high tolerance towards harsh environmental conditions and can live in various environmental niches [28], rendering it a valuable host for treating wastewater pollutants. On the other hand, the ease of preparing various CTS matrices by cross-linking modifications has been established due to its good hydrophilicity, mechanical stability and degree of rigidity [29,30]. The immobilization or embedding of various dye-degrading enzymes or cells with dye-degrading or dye-adsorptive capability on CTS leads to increased decolorization efficiency through the increased pores on the surface of CTS beads [31] or immobilized enzymes/cells [30]. To further improve the performance of our laccase-based cell surface display system in terms of decolorization efficiency, stability and applicability for large-scale or continuous processes, in the present study, using a self-configuring device, a facile and applicable CTS-microsome preparation system was self-assembled and used to prepare five CTS microbead (CTS-MB) materials with different microsome sizes from a pulverous CTS substrate. The CTS-MBs were then used to immobilize engineered *P. putida* MB285 cells with surface-displayed laccase to construct a "CTS-MB/*P. putida* MB285 cell" biosorbent/biodegrading dual-function system. The effects of temperature, pH and storage time on the laccase activity of the optimal complex were investigated. The interactions between the CTS-MBs and *P. putida* cells were examined using scanning electron microscopy (SEM) and Fourier transform infrared (FT-IR) spectroscopy. The material was then used to decolorize five synthetic dyes and its decolorization efficiency together with its capability during 10 rounds of continuous decolorization and 6 cycles of re-culturing were investigated.

2. Materials and Methods

2.1. Chemicals, Bacterial Strains and Culture Conditions

Analytical-grade CTS was purchased from Sinopharm Chemical Reagent Co., Ltd. (Wuhan branch, China). According to the manufacturer's product manual, this off-white translucent powder product is a water-insoluble but dilute acid-soluble polymer with a chemical formula of $(C_6H_{11}NO_4)_n$, an average molecular weight of 6.2×10^5 Da and a degree of deacetylation of $\geq 90.0\%$. Five industrial-grade dyes (Table S1) that belonging to four structural categories were used for the decolorization experiments: azo dyes Acid Red 18 (AR18) and Direct Red 243 (DB243), anthraquinone dye Acid Green 25 (AG25), phthalocyanine dye Reactive Blue 220 (RB220) and tribenyldioxazine dye Reactive Blue 198 (RB198). Among these dyes, AG25 and AR18 were purchased from Thailand Modern Destuffs & Pigments Company (Lardpraw 94, Bangkok, Thailand) and RB198, RB220 and DR243 were purchased from Jinsheng Dyestuff Chemical Co., Ltd. (Jinan, China). Other chemical reagents were purchased from Sinopharm Chemical Reagent Co., Ltd. and were of analytical grade. The recombinant *P. putida* MB285 strain [21] was routinely grown at 28 °C in lysogeny broth (LB) medium [32] containing 500 µg mL^{-1} (final concentration) of carbenicillin.

2.2. Preparation of CTS-MBs

The technical process for preparing the CTS-MBs with surface-grafted aldehyde groups (i.e., the activated CTS-MBs, aCTS-MBs in brief) and the complex with immobilized *P. putida* cells (i.e., the aCTS-MBs complexed with *P. putida* cells, bacCTS-MBs in brief) is illustrated in Figure 1. First, 2 g of CTS powder was dissolved in 100 mL of a 1% (v/v) acetic acid solution, followed by vigorous stirring with a magnetic stirrer for at least 12 h. To prepare CTS-MB with different particle sizes, we designed and self-assembled a set of facile preparation devices consisting of a constant flow pump, a DC high-voltage power supply (HV-PS), a magnetic stirrer and a medical needle (30 G) that was placed at a height of 3.5 cm over a NaOH-ethanol solution (Figure S1). The negative electrode of the HV-PS was connected to a copper ring that was immersed in a NaOH-ethanol solution and the positive electrode of the HV-PS was connected to a needle. CTS-MBs formed in the NaOH-ethanol solution when the CTS solution was titrated from the needle at a flow rate of 0.05 mL min^{-1} and was stirred at 200 rpm. By adjusting the voltage of the HV-PS from 5.3 kV to 0 kV, five CTS-MB preparations with particle diameters ranging from 450 µm to 2100 µm were made. For each, the collected CTS-MBs were washed with double distilled water (ddH$_2$O) to remove NaOH.

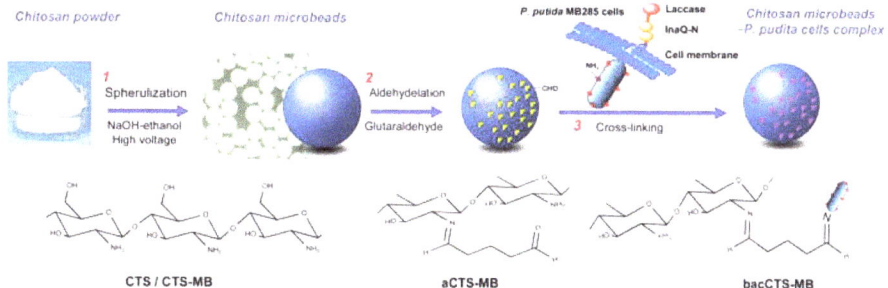

Figure 1. Schematic illustration showing the preparation of the bacCTS-MB complex. 1, Preparation of the CTS-MBs; 2, Surface grafting of aldehyde groups on the CTS-MBs (aCTS-MBs); 3, Immobilization of *P. putida* MB285 cells onto the aCTS-MBs via covalent cross-linkages to prepare the bacCTS-MB materials.

2.3. Surface Aldehyde Modification of the CTS-MBs and Immobilization of P. putida MB285 Cells

To prepare the aCTS-MBs, 2 g of each of the above CTS-MB preparations were dispersed and suspended in 10 mL of 8% glutaraldehyde. Each suspension was shaken for several minutes and modified CTS beads with surface-grafted aldehydic groups were then obtained by centrifugation and washing with ddH$_2$O.

To immobilize *P. putida* cells onto the aCTS-MB materials, an overnight culture of *P. putida* MB285 cells (approximately 1×10^{10} cells mL^{-1}) was harvested, washed three times with sterile phosphate-buffered saline (PBS) (pH 7.4) and diluted to a unit cell density (OD$_{600}$ of 5.0) using PBS buffer (pH 7.4) to make a stock cell suspension. Subsequently, 100 mL of the MB285 cell suspension and an appropriate amount of each aCTS-MB material were added to a 250-mL Erlenmeyer flask (enough to fully immerse the material in the cell suspension). Each mixed suspension was then incubated with agitation at 60 rpm to prepare a complex with CTS-MBs crosslinked by *P. putida* MB285 cells. The crosslinking reaction temperature ranged from 4 °C to 36 °C, the reaction time from 2 h to 12 h and the glutaraldehyde concentrations from 0.5% to 8%. The five prepared CTS-MB materials were subjected to an orthogonal trial (L$_{25}$, 4^5) to optimize the reaction conditions for the immobilization of *P. putida* cells on the CTS-MB materials. The prepared bacCTS-MB materials were washed with sterile PBS buffer (pH 7.4) to remove non-crosslinked cells and residual glutaraldehyde and were then stored at 4 °C until use.

2.4. Assay of IBCC-B Enzymatic Activity

The laccase enzymatic activity of the complexes was measured using 2,2'-azino-bis(3-ethyl-benzthiazoline-6-sulfonic acid) (ABTS) (Amresco) as the substrate at 25 °C, according to a previously described method [22] with minor modifications. Basically, 2 g of each bacCTS-MB complex was added to the reaction system (10-mL total volume), which contained 0.1 mol L^{-1} sodium acetate buffer (pH 2.5), 0.5 mmol L^{-1} ABTS and 2 mmol L^{-1} CuCl$_2$. The ABTS oxidation rate was calculated based on the net increase in absorbance of each reaction mixture at 420 nm using a UV/Vis spectrophotometer (DU-800 Nucleic Acids/Protein Analyzer, Beckman Coulter, Brea, CA, USA). One unit of enzyme activity was defined as the amount of enzyme required to oxidize 1 μmol of ABTS per minute. Each laccase activity assay was performed at least in triplicate.

2.5. Dye Decolorization

Full wavelength (325 nm to 800 nm) scanning of five synthesized dyes was performed using a UV/Vis spectrophotometer to record their maximal adsorbent peaks (Figure S2), which were used as the wavelength values for dye absorbance measurement. Decolorization by bacCTS-MB1 was assessed using a previously described method with some modifications. First, 2 g of each bacCTS-MB1 complex was added to a 10-mL reaction system containing 8.5 mL of 0.1 mol L^{-1} sodium acetate buffer (pH 2.5), 0.5 mL of 2 mmol L^{-1} CuCl$_2$ and 1.0 mL of dye at the final concentration of 1.0 g L^{-1}. Second, the absorbance of each reaction mixture supernatant was spectrophotometrically measured at the maximal absorbance wavelength value of each dye (λ_{max}) (Table S1 and Figure S2). Activity was expressed as the relative decolorization value, which was calculated as follows:

$$\text{Relative decolorization value}(\%) = \frac{(A_0 - A_f)}{A_0} \times 100\% \qquad (1)$$

where A_0 denotes the initial absorbance value and A_f denotes the final absorbance value.

For the ten-round repeated decolorization experiments, 2 g of each prepared bacCTS-MB1 complex was tested for AG25 decolorizing activity (final concentration of 1 g L^{-1} AG25) in shake-flask trials in 10 mL of reaction solution at pH 2.5 and 25 °C with 180 rpm shaking. After each decolorization reaction round, the complex was harvested by centrifugation, repeatedly washed with ddH$_2$O and then directly subjected to the next round of decolorization under similar decolorization reaction condition.

To examine the decolorizing activity of bacCTS-MB1 after a round of decolorization in conjunction with culturing for a generation, the supernatant was removed via centrifugation after a decolorization reaction, repeatedly washed with ddH$_2$O until the supernatant no longer discolored, then 100 mL of LB broth were added directly into the flask, following by incubation at 25 °C for 4 h with 200 rpm of shaking. This procedure was not conducted under strict aseptic conditions. A next-round decolorization reaction was then carried out under similar reaction conditions. A total of six rounds of decolorization with five rounds of culturing for a generation were conducted. The decolorization activity of the bacCTS-MB1 complex with or without culturing for a generation was measured according to the procedures described above and the relative decolorization value of each reaction was calculated according to Formula (1).

2.6. Characterization of the CTS-MBs, aCTS-MBs and bacCTS-MBs

The particle sizes of all prepared CTS-MB materials were measured using a Zetasizer Nano-ZS (Malvern Instruments, Malvern, UK) with 100 randomly selected CTS-MBs from each preparation. The CTS-MB1, aCTS-MB1, bacCTS-MB1 and *P. putida* cell samples were dried using vacuum freeze-drying procedures. These samples were mixed with KBr, pressed into tablets and subjected to FT-IR spectral analyses using an FT-IR spectrometer (Spectrum One; PerkinElmer, Waltham, MA, USA). All infrared spectra were recorded within the 480–4000 cm^{-1} spectral range. For the SEM observations of aCTS-MB1, bacCTS-MB1 and *P. putida* cell morphology, a small amount of each prepared sample was dehydrated using a series of gradient ethanol solutions (40%, 70%, 90% and 100%); subsequently, the samples were vacuum dried, gold coated using a sputter coater and then observed under a JSM-6390/LV SEM (NTC, Tokyo, Japan) following the manufacturer's instructions.

2.7. Data Analysis

Data analysis was performed using SPSS (Statistical Package for the Social Sciences) software, version 17.0. All data presented are the averages of at least three assays. Statistical significance was defined as $P < 0.05$.

3. Results and Discussion

3.1. Preparation and Characterization of the IBCC-B Complexes

Five CTS-MB materials with different particle sizes were prepared from a pulverous CTS substrate using a self-assembled device (Figure S1). With decreasing HV-PS voltage intensity values, the particle size of the prepared CTS-MBs increased continuously. As shown in Figure S3, all prepared CTS-MBs were uniform microspherical beads that were water-insoluble but well-compatible and well-dispersive in water. Among these CTS-MBs, the CTS-MB1 sample, which was prepared at the maximum working voltage (5.3 kV), had the smallest average particle size and largest average specific area compared to those of the other as-prepared materials; conversely, the CTS-MB5 sample prepared at 0 kV had the largest particle size and smallest average specific area.

The engineered *P. putida* cells with surface-displayed laccases were immobilized onto the CTS-MBs through immobilization reactions to construct dual-function biosorbent/biodegrading materials. To increase the immobilization efficiency, the as-prepared CTS-MB materials were initially modified with surface-grafted aldehyde groups, which allow *P. putida* immobilization via covalent crosslinking to the CTS-MBs in addition to electrostatic and physical adsorption. An orthogonal test at four factors/five levels with regard to the effects of glutaraldehyde concentration (0.5%, 1.0%, 2.0%, 5.0% and 8.0%,), reaction time (2, 4, 6, 8 and 12 h), temperature (4, 12, 16, 28 and 36 °C) and CTS-MB preparation (CTS-MB5 to CTS-MB1) was performed to optimize the immobilization reaction conditions based on the whole-complex laccase activity of each bacCTS-MB preparation (Table S2). Table S3 shows that the four factors exhibited an optimized effect on the CTS-MB-initiated immobilization of *P. putida* MB285 in the following order: reaction time > temperature > glutaraldehyde concentration >

CTS-MB particle size. The optimized factors corresponded to the combination "$B_3C_3A_5D_5$" (Table S3), indicating that the optimized treatment conditions were as follows: an immobilization reaction time of 6 h, a temperature of 20 °C and a glutaraldehyde concentration of 8% with the smallest particle-sized CTS-MB1.

Under the above optimized immobilization conditions, complexes bacCTS-MB1s were prepared by immobilizing *P. putida* cells onto CTS-MB1s. For comparison, four other complexes (bacCTS-MB2s, -MB3s, -MB4s and -MB5s) were also prepared in parallel using the CTS-MB2, CTS-MB3, CTS-MB4 and CTS-MB5 matrices, respectively. Figure S4 shows that the bacCTS-MB1s exhibited the highest whole-complex laccase enzyme activity as well as the highest immobilized count of *P. putida* cells, —it is apparently attributable to the smallest specific surface area of CTS-MB1 by which more *P. putida* MB285 cells were bound. Therefore, the bacCTS-MB1 complex was selected as the decolorization material for subsequent experiments.

The immobilization profiles of the bacCTS-MB1 complex, aCTS-MB1s and free *P. putida* MB285 cells were morphologically examined using SEM. The rugged and porous structures of the aCTS-MB1s (Figure 2A) were clearly observed and MB285 cells appeared in a naturally dispersed state (Figure 2B). However, the surface of the bacCTS-MB1 complex was covered by a large number of cells that formed clusters of cell aggregates (Figure 2C). These results indicated that during the formation of "CTS-MB/*P. putida* cells" as an integral composite, the surface-grafted aldehyde groups and porous and ion-charged surfaces of CTS-MB1 caused the immobilization of *P. putida* cells, which constitute an associative biomass.

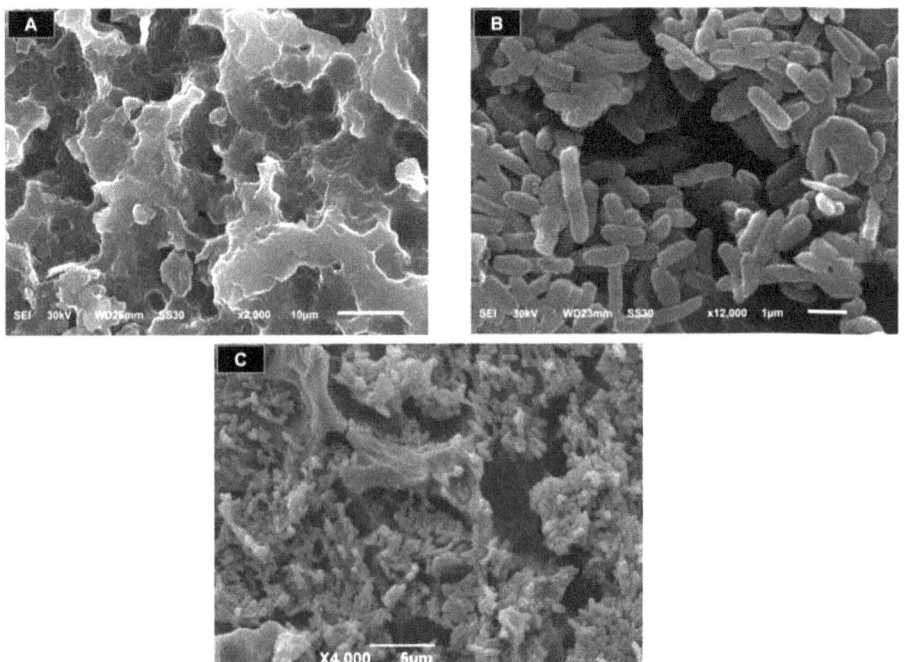

Figure 2. Representative SEM micrograph of the aCTS-MBs (**A**), *P. putida* MB285 cells (**B**) and the bacCTS-MB1 complex (**C**).

FT-IR spectroscopic analyses of free *P. putida* MB285 cells, CTS-MB1s, aCTS-MB1s and bacCTS-MB1s were performed to verify the chemical groups involved in the binding of the CTS-MBs and *P. putida* cells. Figure 3 shows the aCTS-MB1 spectrum in which the displayed peak at 1599 cm^{-1} that represents the δ(N–H) bending vibration absorption peak of the aCTS-MB1s was obviously weakened, whereas the ν(C–N) stretching vibration absorption peak in the 1642–1658 cm^{-1} region was increased significantly; no significant changes were found in the other absorption peaks, suggesting covalent binding between the aldehyde groups of glutaraldehyde and the amino groups. The bacCTS-MB1 spectrum displayed peaks at 1640, 1599 and 1310 cm^{-1}, which represent the ν(C=O) stretching vibration absorption peak, the δ(N–H) bending vibration absorption peak and the ν(C–H) stretching vibration absorption peak, respectively; the latter two peaks were both somewhat weakened, indicating that the acylation in the bacCTS-MB1s was further weakened. Moreover, the ν(O-H) and ν(N–H) stretching vibration absorption peak at 3430–3440 cm^{-1} in the bacCTS-MB1 spectrum was also slightly weakened, which might be due to the binding of amino groups and aldehyde groups on the surface of the bacCTS-MB1s. These results indicated that the variance of surface groups on both the aCTS-MB1s and bacCTS-MB1s that result from surface-grafting modification and *P. putida* cell immobilization might be involved in the binding of cells to the surface of the CTS-MB1s.

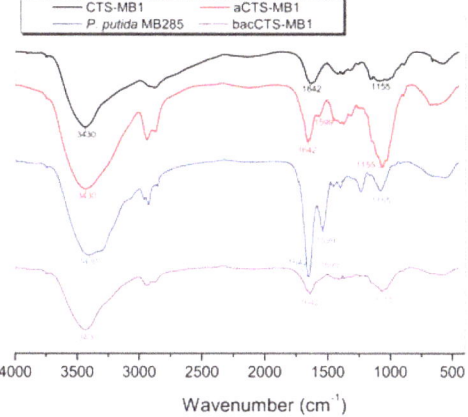

Figure 3. FT-IR spectra for *P. putida* MB285 cells and the as-prepared CTS-MB1s, aCTS-MB1s and bacCTS-MB1s.

3.2. Effects of Temperature, pH and Storage Time on the Laccase Activity of the bacCTS-MB1

The effects of different temperatures on the whole-complex laccase activity of the bacCTS-MB1s and the whole-cell laccase activity of *P. putida* MB285 cells were comparatively examined at a range of 15 °C to 85 °C. As shown in Figure 4A, both the bacCTS-MB1s and MB285 cells exhibited parallel variation profiles along with increasing temperature. From 15 °C to 25 °C, the activity of the bacCTS-MB1s and MB285 cells increased rapidly, with maximum activity observed at 25 °C; however, their activity declined sharply at temperatures above 30 °C. Nevertheless, for the bacCTS-MB1 complex, a greater degree of thermostability was evident, as this enzyme lost only 46.4% of its activity at 55 °C, 52.3% at 65 °C and 83.2% at 75 °C, whereas the MB285 cells lost 57.8% of their enzyme activity at 55 °C, 72.6% at 65 °C and 94.8% at 75 °C. Therefore, the bacCTS-MB1 complex exhibited improved thermostability towards higher temperatures.

Figure 4B shows that although the pH range of the bacCTS-MB1 complex and *P. putida* cells extended from pH 1 to 7, the optimal pH value was 2.5, indicating that the cross-linking immobilization did not alter the optimal pH value of the *P. putida* MB285 laccase. Even at in a pH value range of 3 to 7, the bacCTS-MB1 complex retained more stable laccase activity compared to that of the free *P. putida* MB285 cells, suggesting a sheltering effect of the CTS-MB1 matrix on the immobilized cells, which led to the improved performance of the bacCTS-MB1s at higher pH values.

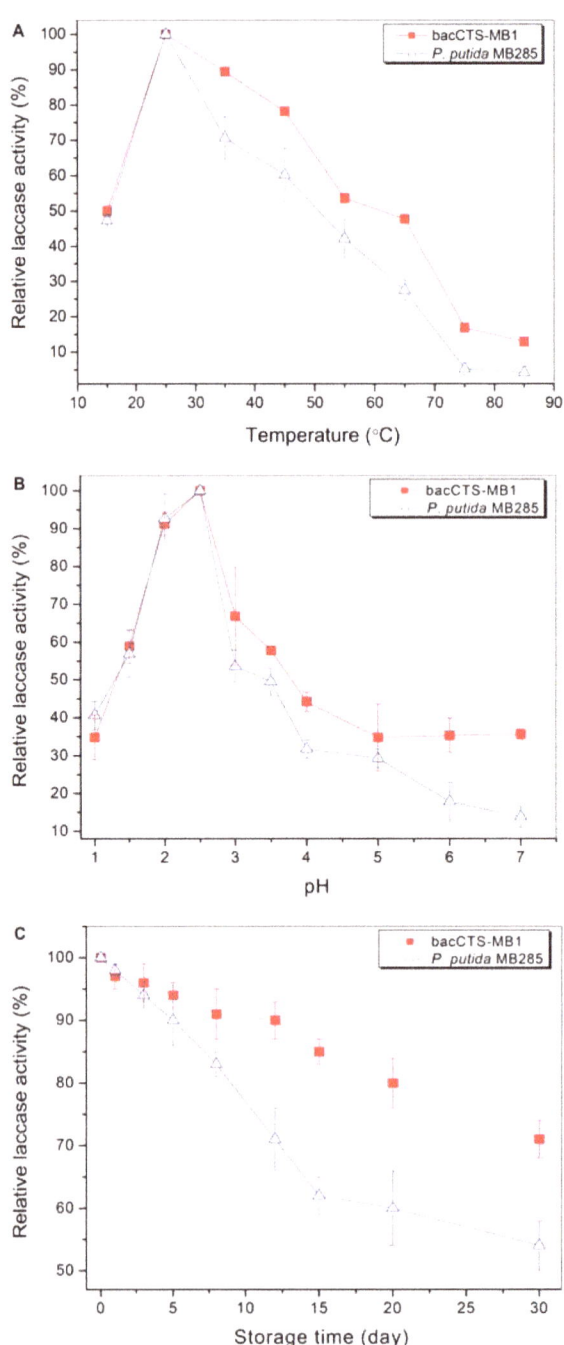

Figure 4. Effect of temperature (**A**), pH value (**B**) and storage time (**C**) on the whole-complex laccase enzyme activity of the bacCTS-MB1 complex. Whole-cell laccase enzyme activity of *P. putida* MB285 cells alone was assayed in parallel.

The bacCTS-MB1 complex and a *P. putida* MB285 cell suspension were stored at 4 °C for 30 days to monitor laccase activity attenuation. As shown in Figure 4C, the bacCTS-MB1 complex was more stable, retaining its enzyme activity over the 30 days of incubation, with the loss of only 15% of its activity at day 15, 20% at day 20 and 29% at day 30; conversely, the free *P. putida* MB285 cells lost 38% of their enzyme activity at day 15, 40% at day 20 and 46% at day 30. These results indicate that a CTS matrix platform is apparently conducive to retain the activity of *P. putida* MB285 cells with surface-displayed laccase enzymes.

3.3. Dye Decolorization by the bacCTS-MB1 Complex

Prior to dye decolorization using the bacCTS-MB1 complex, single factor tests for reaction time, bacCTS-MB1 loading, pH, temperature and shaking speed with 1 g L^{-1} of the representative anthraquinone dye AG25 and the azo dye AR18 were performed to optimize the degradation reaction conditions. The time-course patterns of AG25/AR18 decolorization were initially determined using 2 g of the bacCTS-MB1s under the following reaction conditions: pH 2.5, 25 °C and a shaking speed of 180 rpm (Figure 5A). The results showed that the decolorization of both AG25 and AR18 occurred rapidly, with approximately 90% of the reaction equilibrium value attained within the first 20 min and an increasing trend maintained until 30 min, at which time the final decolorization reaction equilibrium was reached. Therefore, 30 min set as the reaction time for the other factor experiments. Figure 5B shows that the decolorization of AG25 and AR18 using 2 g of the bacCTS-MB1s reached its maximal value in a 30-min reaction. Figure 5C,D show that the bacCTS-MB1s were capable of decolorizing AG25 and AR18 across a wide pH range of 2.5–8.5 and a relatively broad temperature range of 15–85 °C. At an optimal pH value of 2.5 and an optimal temperature of 25 °C, the decolorization reached its maximal value in 30 min. Figure 5E shows that shaking increased the decolorization reaction compared to static conditions, with every shaking speed increasing the relative decolorization rate. Although shaking at 240 rpm exhibited faster decolorization of AG25 compared with that of 180 rpm, both reached similar decolorization values of AG25 and AR18 in 30 min. Thus, the optimized decolorization reaction conditions were defined as follows: a reaction time of 30 min with a loading of 2 g of the bacCTS-MB1s at a pH of 2.5, a temperature of 25 °C and a shaking speed of 180 rpm.

The decolorization capacity of the bacCTS-MB1s on five industrial dyes was determined under optimized shake-flask incubation conditions. Figure 6 shows that the bacCTS-MB1 complex exhibited remarkable dye decolorization, with a maximum relative decolorization value of 96.2% for AG25, 95.6% for AR18, 96.3% for RB198, 95.7% for RB220 and 94.3% for DR243 in 30 min (Figure 6A); conversely, the cell-free CTS-MBs exhibited limited decolorization capacity with 29.0%, 28.7%, 16.2%, 36.0% and 21.9% decolorization values, respectively (Figure 6B). These results indicated that the immobilized *P. putida* MB285 cells contributed to the decolorization of dyes and the CTS-MB matrix subsidiarity increased the total decolorization capacity via its biosorption activity.

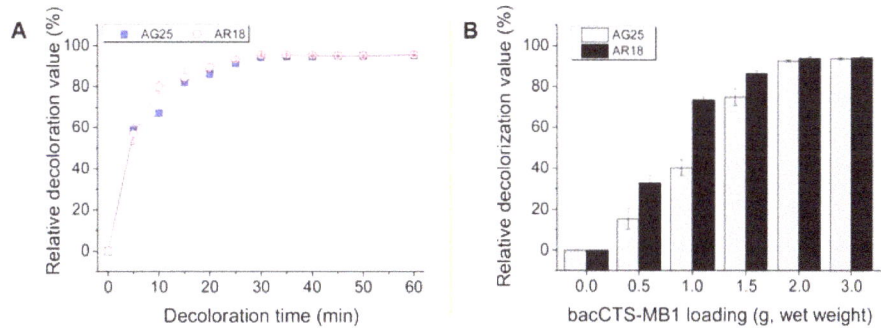

Figure 5. *Cont.*

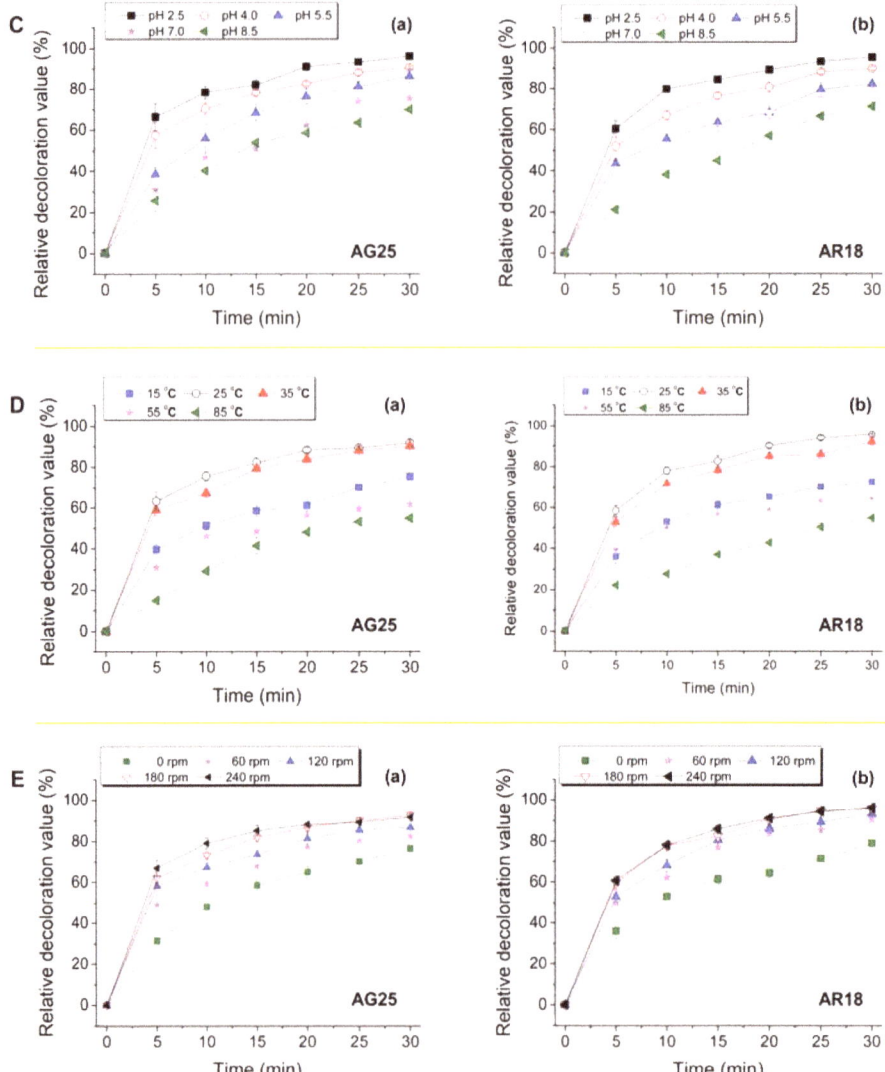

Figure 5. Effect of reaction time (**A**), bacCTS-MB1 loading (**B**), pH value (**C**), temperature (**D**) and shaking speed (**E**) on AG25 and AR18 decolorization by the IBCC-B1s. Each reaction included 2 g (unless otherwise specified) of the bacCTS-MB1s (wet weight) and a final concentration of 1 g L^{-1} of the dye substrate.

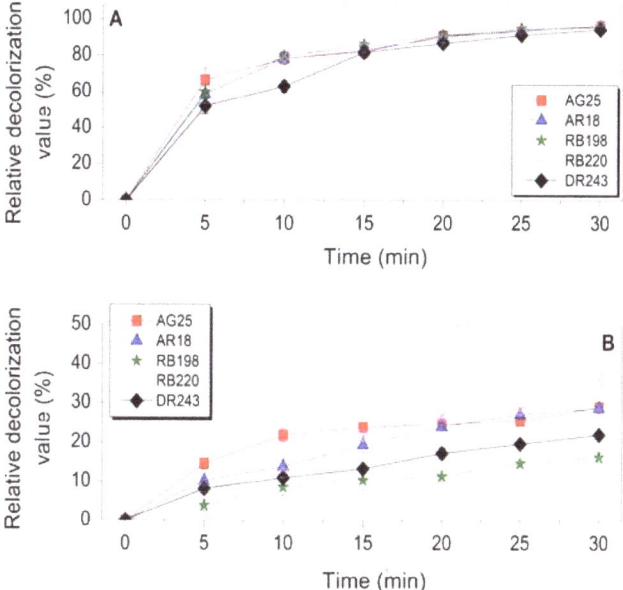

Figure 6. Dye decolorization of bacCTS-MB1s (**A**) and CTS-MB1s (**B**) towards AG25, AR18, RB198, RB220 and DR243. Decolorization was performed using 2 g of the bacCTS-MB1s or CTS-MB1s (wet weight) at 28 °C and pH 2.5 with shaking at 180 rpm for 30 min.

3.4. Effect of Repeated Use on AG25 Decolorization by the bacCTS-MB1s

Figure 7 shows that the bacCTS-MB1 complex maintained substantial AG25 decolorization capacity with repeated use, retaining over 80% of its decolorization value after the first five rounds of repeated reactions, which represents a loss of only 13% in the 5th round compared to the first round. Although relatively rapid loss occurred from the 6th to the 10th round, the complex still retained a relative decolorization value of over 40%. These results reflected the strong viability of *P. putida* MB285 cells under such treatments, thereby suggesting that the bacCTS-MB1s have good persistent efficacy during decolorization of the AG25 dye.

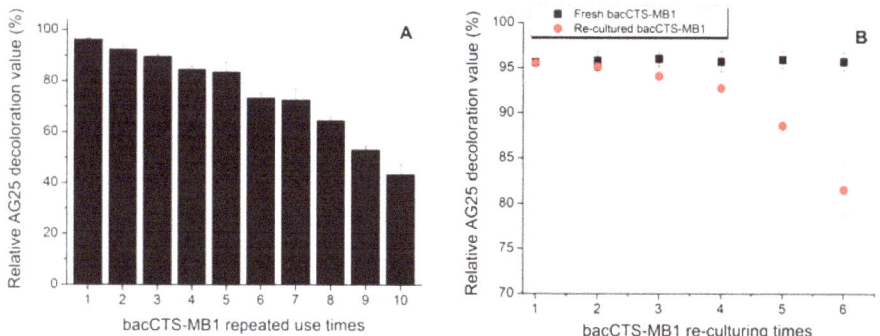

Figure 7. AG25 decolorization capacity of the bacCTS-MB1s during continuously repeated reactions (**A**) and during continuous decolorization and re-culturing cycles (**B**). In (**B**), equivalent freshly prepared bacCTS-MB1s were used as a control for each decolorization reaction cycle.

3.5. Effect of Re-Culturing Time on Decolorization by the bacCTS-MB1s

The re-culture and reusability of a dye-decolorizing material affects its potential for continuous decolorization processes. Six continuous rounds of AG25 decolorization and with re-culture were performed via shake flask trials to examine the effect of re-culture on the decolorization capacity of the bacCTS-MB1s. Figure 7B shows that the first four rounds of re-cultured bacCTS-MB1s had a decolorization value approximately 8% greater than that of the directly used bacCTS-MB1 complex (Figure 7A), indicating that though effective, the re-culturing process was inappropriate during the first 3-4 rounds of use for the purpose of facile and fast material retrieval.

4. Conclusions

The present study reports a new decolorization material that is capable of decolorizing different kinds of dyes with high capacity and improved physical strength for potential applications in large-scale or continuous operations. Following the preparation of CTS-MBs using a self-configuring device, engineered *P. putida* MB285 cells with surface-displayed laccases were immobilized mainly through covalent cross-linkages, thus forming a degrading-biosorption bifunctional complex. SEM and infrared analysis confirmed the successful immobilization of the cells onto the CTS-MB matrix. The prepared complexes with maximum whole-complex enzyme activity, bacCTS-MB1, was used to decolorize five industrially used dyes in shake flask trials. The results showed that this complex exhibited high decolorization capacity under the optimized reaction conditions. Moreover, the bacCTS-MB1 complex showed favorable activity with repeated and regenerated use. Therefore, the bacCTS-MB1 complex efficiently immobilized *P. putida* cells and demonstrated enhanced decolorization activity and improved physical strength. This system could be potential for further applications in large-scale or continuous processes; however, the feasibility of this system must be validated in naturally occurring, polluted bodies of water where various dyes and other pollutants coexist and could cause interference. The development of an applicable system based on the bacCTS-MB1s to treat industrial dye-polluted waters will be our next primary research goal.

Supplementary Materials: The following are available online at http://www.mdpi.com/2076-3417/9/1/138/s1, Figure S1: Schematic illustration of the self-assembled facile device for the preparation of various CTS-MBs with different particle sizes; Figure S2: Morphological images of the five as-prepared CTS-MB materials and their average diameters and specific surface area values; Figure S3: Full wavelength (325 nm to 800 nm) scanning curves of five synthesized dyes indicating their maximal adsorbent peaks (arrows), which were used as the OD values for the dye absorbance measurements; Figure S4: Whole-composite laccase enzyme activity and immobilized *P. putida* cell counts of the five prepared bacCTS-MB materials; Table S1: Molecular formulas of five selected synthesized dyes; Table S2: L_{25} (4^5)-orthogonal test of *P. putida* MB285 immobilization on the CTS-MBs; Table S3: Significance analysis of the factors in the L_{25}-orthogonal test of *P. putida* MB285-immobilization.

Author Contributions: Z.B., X.S. and X.Y. performed the experiments. Z.B. drafted the manuscript. L.L. conceived and directed the study and revised the manuscript.

Funding: This work was funded by the National Natural Science Foundation of China (Grant No. 31570123 and 31770108) and the Fundamental Research Funds for the Central Universities (Program No. 2662015PY189).

Conflicts of Interest: The authors declare no conflict of interest.

References

1. Husain, Q. Potential applications of the oxidoreductive enzymes in the decolorization and detoxification of textile and other synthetic dyes from polluted water: A review. *Crit. Rev. Biotechnol.* **2006**, *26*, 201–221. [CrossRef] [PubMed]

2. Srinivasan, A.; Viraraghavan, T. Decolorization of dye wastewaters by biosorbents: A review. *J. Environ. Manag.* **2010**, *91*, 1915–1929. [CrossRef] [PubMed]

3. Cheng, S. Heavy metal pollution in China: Origin, pattern and control. *Environ. Sci. Pollut. Res. Int.* **2003**, *10*, 192–198. [CrossRef]

4.   Dos Santos, A.B.; Cervantes, F.J.; van Lier, J.B. Review paper on current technologies for decolourisation of textile wastewaters: Perspectives for anaerobic biotechnology. *Bioresour. Technol.* **2007**, *98*, 2369–2385. [CrossRef] [PubMed]

5.   Nidheesh, P.V.; Gandhimathi, R.; Ramesh, S.T. Degradation of dyes from aqueous solution by Fenton processes: A review. *Environ. Sci. Pollut. Res. Int.* **2013**, *20*, 2099–2132. [CrossRef] [PubMed]

6.   Muxika, A.; Etxabide, A.; Uranga, J.; Guerrero, P.; de la Caba, K. Chitosan as a bioactive polymer: Processing, properties and applications. *Int. J. Biol. Macromol.* **2017**, *105*, 1358–1368. [CrossRef]

7.   Farzana, M.H.; Meenakshi, S. Photo-decolorization and detoxification of toxic dyes using titanium dioxide impregnated chitosan beads. *Int. J. Biol. Macromol.* **2014**, *70*, 420–426. [CrossRef]

8.   Bilal, M.; Asgher, M.; Iqbal, M.; Hu, H.; Zhang, X. Chitosan beads immobilized manganese peroxidase catalytic potential for detoxification and decolorization of textile effluent. *Int. J. Biol. Macromol.* **2016**, *89*, 181–189. [CrossRef]

9.   Lin, Y.H.; Lin, W.F.; Jhang, K.N.; Lin, P.Y.; Lee, M.C. Adsorption with biodegradation for decolorization of reactive black 5 by Funalia trogii 200800 on a fly ash-chitosan medium in a fluidized bed bioreactor-kinetic model and reactor performance. *Biodegradation* **2013**, *24*, 137–152. [CrossRef]

10. Wang, J.; Chen, C. Chitosan-based biosorbents, modification and application for biosorption of heavy metals and radionuclides. *Bioresour. Technol.* **2014**, *160*, 129–141. [CrossRef]

11. Philibert, T.; Lee, B.H.; Fabien, N. Current status and new perspectives on chitin and chitosan as functional biopolymers. *Appl. Biochem. Biotechnol.* **2017**, *181*, 1314–1337. [CrossRef] [PubMed]

12. Thakur, V.K.; Voicu, S.I. Recent advances in cellulose and chitosan based membranes for water purification: A concise review. *Carbohydr. Polym.* **2016**, *146*, 148–165. [CrossRef]

13. Yang, R.; Li, H.; Huang, M.; Yang, H.; Li, A. A review on chitosan-based flocculants and their applications in water treatment. *Water Res.* **2016**, *95*, 59–89. [CrossRef]

14. LogithKumar, R.; KeshavNarayan, A.; Dhivya, S.; Chawla, A.; Saravanan, S.; Selvamurugan, N. A review of chitosan and its derivatives in bone tissue engineering. *Carbohydr. Polym.* **2016**, *151*, 172–188. [CrossRef] [PubMed]

15. Muanprasat, C.; Chatsudthipong, V. Chitosan oligosaccharide: Biological activities and potential therapeutic applications. *Pharmacol. Ther.* **2017**, *170*, 80–97. [CrossRef] [PubMed]

16. Canas, A.I.; Camarero, S. Laccases and their natural mediators: Biotechnological tools for sustainable eco-friendly processes. *Biotechnol. Adv.* **2010**, *28*, 694–705. [CrossRef] [PubMed]

17. Ba, S.; Vinoth Kumar, V. Recent developments in the use of tyrosinase and laccase in environmental applications. *Crit. Rev. Biotechnol.* **2017**, *37*, 819–832. [CrossRef] [PubMed]

18. Claus, H. Laccases and their occurrence in prokaryotes. *Arch. Microbiol.* **2003**, *179*, 145–150. [CrossRef]

19. Claus, H. Laccases: Structure, reactions, distribution. *Micron* **2004**, *35*, 93–96. [CrossRef]

20. Bertrand, B.; Martinez-Morales, F.; Trejo-Hernandez, M.R. Upgrading laccase production and biochemical properties: Strategies and challenges. *Biotechnol. Prog.* **2017**, *33*, 1015–1034. [CrossRef]

21. Chauhan, P.S.; Goradia, B.; Saxena, A. Bacterial laccase: Recent update on production, properties and industrial applications. *3 Biotech* **2017**, *7*, 323. [CrossRef] [PubMed]

22. Chen, Y.; Stemple, B.; Kumar, M.; Wei, N. Cell surface display fungal laccase as a renewable biocatalyst for degradation of persistent micropollutants bisphenol A and sulfamethoxazole. *Environ. Sci. Technol.* **2016**, *50*, 8799–8808. [CrossRef] [PubMed]

23. Couto, S.R.; Toca-Herrera, J.L. Industrial and biotechnological applications of laccass: A review. *Biotechnol. Adv.* **2006**, *24*, 500–513. [CrossRef] [PubMed]

24. Peralta-Zamora, P.; Pereira, C.M.; Tiburtius, E.R.L.; Moraes, S.G.; Rosa, M.A.; Minussi, R.C.; Durán, N. Decolorization of reactive dyes by immobilized laccase. *Appl. Catal. B Environ.* **2003**, *42*, 131–144. [CrossRef]

25. Wang, W.; Zhang, Z.; Ni, H.; Yang, X.; Li, Q.; Li, L. Decolorization of industrial synthetic dyes using engineered *Pseudomonas putida* cells with surface-immobilized bacterial laccase. *Microb. Cell Fact.* **2012**, *11*, 75. [CrossRef] [PubMed]

26. Shao, X.; Gao, Y.; Jiang, M.; Li, L. Deletion and site-directed mutagenesis of laccase from *Shigella dysenteriae* results in enhanced enzymatic activity and thermostability. *Enzyme Microb. Technol.* **2009**, *44*, 274–280. [CrossRef]

27. Ramos, J.L.; Sol Cuenca, M.; Molina-Santiago, C.; Segura, A.; Duque, E.; Gomez-Garcia, M.R.; Udaondo, Z.; Roca, A. Mechanisms of solvent resistance mediated by interplay of cellular factors in *Pseudomonas putida*. *FEMS Microbiol. Rev.* **2015**, *39*, 555–566. [CrossRef]

28. Kim, J.; Park, W. Oxidative stress response in *Pseudomonas putida*. *Appl. Microbiol. Biotechnol.* **2014**, *98*, 6933–6946. [CrossRef]

29. El Kadib, A. Chitosan as a sustainable organocatalyst: A concise overview. *ChemSusChem* **2015**, *8*, 217–244. [CrossRef]

30. Krajewska, B. Application of chitin- and chitosan-based materials for enzyme immobilizations: A review. *Enzyme Microb. Technol.* **2004**, *35*, 126–139. [CrossRef]

31. Chiou, M.S.; Chuang, G.S. Competitive adsorption of dye metanil yellow and RB15 in acid solutions on chemically cross-linked chitosan beads. *Chemosphere* **2006**, *62*, 731–740. [CrossRef] [PubMed]

32. Sambrook, J.; Russell, D.W. *Molecular Cloning: A Laboratory Manual*, 3rd ed.; Cold Spring Harbor Laboratory Press: Cold Spring Harbor, NY, USA, 2001.

MDPI

St. Alban-Anlage 66

4052 Basel

Switzerland

Tel. +41 61 683 77 34

Fax +41 61 302 89 18

www.mdpi.com

Applied Sciences Editorial Office

E-mail: applsci@mdpi.com

www.mdpi.com/journal/applsci

CPSIA information can be obtained
at www.ICGtesting.com
Printed in the USA
BVHW020824280819

556820BV00023BA/667/P

MDPI
St. Alban-Anlage 66
4052 Basel
Switzerland

Tel: +41 61 683 77 34
Fax: +41 61 302 89 18

www.mdpi.com

ISBN 978-3-03921-367-2